Llewellynn Frederick William Jewitt

Grave-Mounds and Their Contents

A manual of archaeology, as exemplified in the burials of the Celtic, the Romano-British, and the Anglo-Saxon periods

Llewellynn Frederick William Jewitt

Grave-Mounds and Their Contents

A manual of archaeology, as exemplified in the burials of the Celtic, the Romano-British, and the Anglo-Saxon periods

ISBN/EAN: 9783337019570

Printed in Europe, USA, Canada, Australia, Japan

Cover: Foto ©berggeist007 / pixelio.de

More available books at **www.hansebooks.com**

GRAVE-MOUNDS AND THEIR CONTENTS.

Grave-mounds and their Contents:

A MANUAL OF ARCHÆOLOGY,

AS EXEMPLIFIED IN THE

BURIALS OF THE CELTIC, THE ROMANO-BRITISH, AND THE ANGLO-SAXON PERIODS.

BY

LLEWELLYNN JEWITT, F.S.A.,

ETC., ETC.

With nearly Five Hundred Illustrations.

LONDON:
GROOMBRIDGE AND SONS,
5, PATERNOSTER ROW.

1870.

TO

MY OLD AND MUCH-ESTEEMED FRIEND,

JOSEPH MAYER, Esq.,

FELLOW OF THE SOCIETY OF ANTIQUARIES OF LONDON;

FELLOW OF THE ROYAL SOCIETY OF NORTHERN ANTIQUARIES

OF COPENHAGEN;

FELLOW OF THE ROYAL ASIATIC SOCIETY;

MEMBER OF THE SOCIETIES OF ANTIQUARIES OF

FRANCE, NORMANDY, THE MORINI,

ETC., ETC., ETC.;

ONE OF THE MOST ARDENT AND ZEALOUS OF ARCHÆOLOGISTS,

AND MOST KINDLY OF MEN;

THE PRINCELY DONOR TO THE PUBLIC

OF THE FINEST AND MOST EXTENSIVE

MUSEUM OF ANTIQUITIES

EVER COLLECTED TOGETHER BY A SINGLE INDIVIDUAL;

I, WITH TRUE PLEASURE,

𝔇𝔢𝔡𝔦𝔠𝔞𝔱𝔢 𝔱𝔥𝔦𝔰 𝔙𝔬𝔩𝔲𝔪𝔢.

LLEWELLYNN JEWITT.

CONTENTS.

CHAPTER I.

Grave-mounds in general—Their Historical Importance—General Situation—Known as Barrows, Houes, Tumps, and Lows—List of Names—Division into Periods 1

CHAPTER II.

Ancient British or Celtic Period—General characteristics of the Barrows—Modes of construction—Interments by inhumation and by cremation—Positions of the Body—Hitter Hill Barrow—Elliptical Barrow at Swinscoe—Burial in contracted position—In sitting and kneeling positions—Double Interments 6

CHAPTER III.

Ancient British or Celtic Period—Interment by cremation—Discovery of lead—Burial in Urns—Positions of Urns—Heaps of burnt Bones—Burnt Bones enclosed in cloth and skins—Stone Cists—Long-Low—Liff's-Low, etc.—Pit Interments—Tree-coffins 31

CHAPTER IV.

Ancient British or Celtic Period—Sepulchral Chambers of Stone—Cromlechs—Chambered Tumuli—New Grange and Dowth—The Channel Islands—Wieland Smith's Cave, and others—Stone Circles—For what purpose formed—Formation of Grave-mounds—Varieties of Stone Circles—Examples of different kinds—Arbor-Low, etc. 50

CHAPTER V.

Ancient British or Celtic Period—Pottery—Mode of manufacture—Arrangement in classes—Cinerary or Sepulchral Urns—Food Vessels—Drinking-cups—Incense Cups—Probably Sepulchral Urns for Infants—Other examples of Pottery . . . 83

CHAPTER VI.

Ancient British or Celtic Period—Implements of Stone—Celts—Stone Hammers—Stone Hatchets, Mauls, etc.—Triturating Stones—Flint Implements—Classification of Flints—Jet articles—Necklaces, Studs, etc.—Bone Instruments—Bronze Celts, Daggers, etc.—Gold articles 109

CHAPTER VII.

Romano-British Period—General Characteristics—Modes of Burial—Customs attendant on Burial—Interments by cremation and by inhumation—Barrows—Tombs of Stone—Lead Coffins—Clay and Tile Coffins—Sepulchral Inscriptions, etc. . . 134

CHAPTER VIII.

Romano-British Period—Pottery—Durobrivian Ware—Upchurch Ware—Salopian Ware—Pottery found at Uriconium—Potteries of the New Forest, of Yorkshire, and of other places—Sepulchral Urns—Domestic and other vessels . . . 151

CHAPTER IX.

Romano-British Period—Pottery—Samian Ware—Potters' Stamps—Varieties of Ornamentation—Glass Vessels—Sepulchral Vases, etc.—Lachrymatories—Bowls—Beads—Coins found with Interments 175

CHAPTER X.

Romano-British Period—Arms—Swords—Spears, etc.—Knives—Fibulæ—Armillæ—Torques of Gold, etc.—Other Personal Ornaments 190

CHAPTER XI.

Anglo-Saxon Period—Distribution of Anglo-Saxon Population over England—General characteristics of Grave-mounds—Modes of Burial—Poem of Beowulf—Interments by cremation and by inhumation—Articles deposited with the Dead—Positions of the Body—Double and other Interments—Burial in Urns—Cemeteries and Barrows 202

CHAPTER XII.

Anglo-Saxon Period—Pottery, general characteristics of—Cinerary Urns—Saxon Urn with Roman Inscription—Frankish and other Urns—Cemeteries at Kings Newton, etc.—Mode of manufacture—Impressed Ornaments 214

CHAPTER XIII.

Anglo-Saxon Period—Glass Vessels—Drinking-glasses—Tumblers—Ale-glasses—Beads—Necklaces—Ear-rings—Coins, etc. . 228

CHAPTER XIV.

Anglo-Saxon Period—Arms—Swords—Knives—Spears—Shields—Umbones of Shields—Buckles—Helmets—Benty-Grange Tumulus—The Sacred Boar—Grave at Barlaston—Enamelled Discs and pendant Ornaments, etc.—Horse-shoes . . 236

CHAPTER XV.

Anglo-Saxon Period—Fibulæ—Enamelled circular Fibulæ—Gold Fibulæ—Pendant Cross—Cruciform Fibulæ—Penannular Fibulæ—Irish and English examples—Pendant Ornaments, etc. . 266

CHAPTER XVI.

Anglo-Saxon Period—Buckets—Drinking-cups of wood—Bronze Bowls—Bronze Boxes—Combs—Tweezers—Châtelaines—Girdle Ornaments—Keys—Hair-pins—Counters, or Draughtmen, and Dice—Querns—Triturating Stones, etc.—Conclusion 280

LIST OF ILLUSTRATIONS.

ANCIENT BRITISH OR CELTIC PERIOD.

Fig.		Page
1	Gib Hill Tumulus. (Frontispiece.)	
2	Section of Grave-mound, Lord's Down, Dewlish, Dorset	8
3	,, ,, Gib Hill, Derbyshire	9
4	,, ,, with two modes of interment by cremation	10
5	,, ,, with inverted urn	10
6	Interment, Smerril Moor	12
7	,, Tissington	13
8	,, Roundway Hill	14
9	,, Hitter Hill	15
10	Plan of Barrow, showing interments, Hitter Hill	17
11	Interments, Hitter Hill	18
12	,, Hitter Hill	20
13	Skull, Hitter Hill	21
14	Plan, with interments, Swinscoe	23
15	Interment, sitting position, Parcelly Hay	26
16	,, ,, Monsal Dale	28
17	Plan of Barrow, with interments, Monsal Dale	29
18	,, ,, Cressbrook	30
19	Section of Barrow	32
20	Stone Cist, Baslow Moor	33
21	Section of Barrow	33
22	Inverted Urn, with burnt bones, Wardlow	34
23	Plan of Long-Low	36
24	Stone Cist, Long-Low	38
25, 26	Skull, Long-Low	39

LIST OF ILLUSTRATIONS.

Fig.			Page
27	Mode of construction, Long-Low		40
28	Stone Cist, Middleton Moor		41
29	,, Liff's-Low		42
30	,, Gib Hill		43
31	Portion of woollen garment, Scale House		45
32, 33	Tree-coffin, Gristhorpe		48
34	Dagger, etc., Gristhorpe		49
35	Flint, etc., Gristhorpe		49
36	Tree-coffin, boat shape		49
37	Cromlech, Lanyon		52
38	,, ,, plan		52
39	,, Chun		53
40	,, ,, plan		53
41	,, Plas Newydd		54
42	Minning-Low, plan		55
43	New Grange, general view		56
44	,, entrance to passage		57
45	Dowth, general view		58
46	,, entrance to passage		58
47	Cromlech, Knockeen		59
48	,, ,, plan		59
49	,, Gaulstown		60
50	,, ,, plan		60
51	,, Ballynageerah		61
52	,, ,,		62
53	,, ,, plan		62
54	,, L'Ancresse		63
55	Chambered Tumulus, Gavr Innis, south entrance		64
56	,, ,, opening in north chamber		65
57	,, ,, plan of chamber		65
58	,, ,, carved stones		65
59	,, ,, ,,		66
60	,, ,, ,,		66
61	,, ,, ,,		67
62	,, Stony Littleton		68
63	,, ,, plan		58
64	,, Five Wells, Taddington		69
65	,, ,, plan		69
66	Flax Dale Barrow, plan		72
67	Section, Elk-Low		73
68	Stone Circle, "Nine Ladies," Stanton Moor		73
69	,, ,, ,, plan		74

Fig.							Page
70	Stone Circle, construction of		.	.	.	.	75
71	,,	Sancreed	.	.	.	.	76
72	,,	.	.	.	.	.	76
73	,,	Isle of Man	.	.	.	.	76
74	,,	Trewavas Head	.	.	.	.	77
75	,,	Mule Hill	.	.	.	.	77
76	,,	Channel Islands	.	.	.	.	78
77	,,	with "twin barrow"	.	.	.	.	78
78	,,	,, ,,	.	.	.	.	79
79	,,	construction	.	.	.	.	79
80	,,	Boscawen-un	.	.	.	.	80
81	,,	Aber	.	.	.	.	80
82	,,	Berriew	.	.	.	.	81
83	,,	Penmeanmaur	.	.	.	.	81
84	,,	Arbor-Low, Derbyshire	.	.	.	.	82
85	Pottery—fragment, Darwen		.	.	.	.	86
86	,,	Cinerary Urn, Monsal Dale	.	.	.	.	87
87	,,	,,	Ballidon Moor	.	.	.	88
88	,,	,,	,, *in situ*	.	.	.	88
89	,,	,,	Trentham	.	.	.	89
90	,,	,,	Darwen	.	.	.	90
91	,,	,,	Dorsetshire	.	.	.	91
92	,,	,,	,,	.	.	.	91
93	,,	,,	Darley Dale	.	.	.	92
94	,,	,,	Stone	.	.	.	93
95	,,	,,	Cleatham	.	.	.	93
96	,,	,,	,,	.	.	.	93
97	,,	,,	Calais Wold, Yorkshire	.	.	.	94
98	,,	,,	Darley Dale	.	.	.	95
99	,,	,,	Tredinney	.	.	.	96
100	,,	,,	Morvah	.	.	.	96
101	,,	Food vessel, Trentham		.	.	.	97
102	,,	,,	Fimber	.	.	.	98
103	,,	,,	Hitter Hill	.	.	.	98
104	,,	,,	,,	.	.	.	99
105	,,	,,	Monsal Dale	.	.	.	100
106	,,	,,	Fimber	.	.	.	100
107	,,	Drinking-cup, Fimber		.	.	.	102
108	,,	,,	Hay Top, Monsal Dale	.	.	.	102
109	,,	,,	Grind-Low	.	.	.	102
110	,,	,,	Elk-Low	.	.	.	103
111	,,	,,	,, bottom	.	.	.	104

xiv   LIST OF ILLUSTRATIONS.

Fig.		Page
112	Pottery—Drinking-cup, Roundway Hill	104
113	,, ,, Gospel Hillock	105
114	,, Incense cup, Yorkshire	106
115	,, ,, ,,	106
116	,, ,, ,,	106
117	,, ,, ,,	106
118	,, ,, ,,	106
119	,, ,, ,,	106
120	,, ,, Dorsetshire	106
121	,, ,, Yorkshire	106
122	,, ,, Stanton Moor	107
123	,, ,, Baslow	107
124	,, ,, Dorsetshire	107
125	,, ,, Darley Dale	107
126	,, Handled vessel, Yorkshire	108
127, 128, 129, 130, 131	Stone celts, Royal Irish Academy	110
132	Stone celt, Derbyshire example	110
133	Stone hammer-head, Wollaton	111
134	Stone celt, Derbyshire example	112
135	Stone hammer-head, Winster	112
136	,, Cambridgeshire	112
137	,, Trentham	112
138	,, Dorsetshire	113
139, 140, 141	Stone mauls, Royal Irish Academy	113
142	Stone hammer-head, Mickleover	113
143, 144	Triturating stones	114
145	Spindle whorl	115
146, 147, 148, 149, 150, 151, 152, 153	Flint arrow-heads, Greenlow and other places in Derbyshire	116

LIST OF ILLUSTRATIONS.

Fig.			Page
154	Flint, Green-Low		117
155	,, Arbor-Low		118
156	,, Calais Wold		119
157	,, ,,		119
158	,, Gunthorpe		119
159	,, Ringham-Low		119
160	,, Calais Wold		119
161	,, ,,		119
162–169	,, Derbyshire examples		120—122
170	Flint celt, Gospel Hillock		122
171	Necklace of jet, Middleton Moor		123
172	,, jet and bone		124
173	,, jet, Fimber		125
174	Jet studs, Gospel Hillock		126
175	,, Calais Wold		126
176	Jet pendant, Derbyshire		126
177	Bone implement, Green-Low		126
178–181	,, Thors Cave, etc.		127
182	Bone pendant, Arbor-Low		127
183	Bronze celt, Royal Irish Academy		128
184	,, ,,		129
185	,, ,,		129
186	,, ,,		129
187	,, Moot-Low		129
188	,, Royal Irish Academy		131
189	,, ,,		131
190	,, ,,		131
191	,, ,,		131
192	,, ,,		131
193	,, ,,		131
194	,, ,,		131
195	,, ,,		131

Fig.			Page
196	Bronze celt, Royal Irish Academy	. .	131
197	Bronze socketed celt, Kirk Ireton	. .	131
198	Bronze dagger, Bottisham	. . .	132
199	Coin, Mount Batten .	. . .	133
200	,, ,,	. . .	133
201	,, ,,	. . .	133
202	,, ,,	. . .	133
203	,, ,,	. . .	133
204	,, Birkhill	. . .	133
205	,, Blandford	. . .	133

ROMANO-BRITISH PERIOD.

206	Cist of stone, York .	. . .	144
207	Leaden coffin, Colchester	. . .	144
208	,, Bishopstoke	. . .	145
209	Clay coffin, Aldborough	. . .	145
210	Chest of stone, with pottery, etc., Avisford	.	147
211	Tomb of tiles, York .	. . .	148
212	Potter's kiln, Castor .	. . .	152
213) 214 } Pottery, Headington 215)		.	154
216	,, Durobrivian ware, scroll ornaments	.	155
217	,, ,, ,,	. .	157
218	,, ,, ,,	. .	157
219	,, ,, ,,	.	157
220	,, ,, hare hunting	. .	157
221	,, ,, ,,	. .	158
222	,, ,, ,,	. .	158
223	,, ,, stag hunting .	. .	158
224	,, ,, ,, .	. .	158
225	,, ,, ,,	. .	159
226	,, ,, indented urn.	. .	161
227	,, ,, cup .	. .	161
228	,, ,, indented urn	. .	161
229	,, ,, ,,	. .	161
230	,, Upchurch ware, group of vessels	. .	163
231	,, ,, urn	. .	164
232	,, Salopian ware, group of vessels	. .	166
233	,, group, Headington	. .	167
234	,, Sepulchral Urn, Toot Hill	. .	167

LIST OF ILLUSTRATIONS.

Fig.						Page
235	Pottery, Sepulchral Urn, Little Chester					168
236	,,	,,	Leicester			168
237	,,	,,	Little Chester			169
238	,,	,,				169
239	,,	,,				169
240	,,	,,	London			170
241	,,	,,	Water Newton			170
242	,,	,,	London			170
243	,,	,,	Leicester			170
244	,,	,,	Winchester			170
245	,,	,,	Castor			170
246	,,	,,	Water Newton			170
247	,,	,,	Castor			170
248	,,	,,	Castor			171
249	,,	,,	London			171
250	,,	Amphora	,,			172
251	,,	,,	,,			172
252	,,	,,	,,			172
253	,,	,,	,,			172
254	,,	Mortarium	,,			171
255	,,	group, Headington				173
256	,,	Headington				173
357	,,	,,				173
258	,,	,,				173
259–266	,,	various localities				174
267	,,	Samian ware, cup, Leicester				175
268 to 275	Potters' marks					176
276	Potters' stamp					177
277	Pottery, Samian ware, bowl, London					178
278	,,	,,	,,	Arezzo		179
279	,,	,,	,,	,,		179
280	,,	,,	,,	London		179
281	,,	,,	,,	,,		180
282	,,	,,	patera	,,		180
283	,,	,,	foliage			181

Fig.		Page
284 Pottery, Samian ware, bowl, Castor	. . .	182
285 ,, ,, bacchanalian scene	. . .	182
286 ,, ,, bowl, Castor	. . .	183
287 ,, ,, ,, Cologne	. . .	183
288 to 291 ,, cups, etc., London	. . .	184
292 Glass, sepulchral vessel, Bartlow Hills	. . .	185
293 ,, bowl, Leicester	. . .	186
294 to 297 Glass beads	. . .	186
298 Glass beads, etc.	. . .	187
299 Sword, Lough Lea	. . .	191
300 ,, Lincolnshire	. . .	191
301 Spear or lance-head, Little Chester	. . .	190
302 ,, ,, Harnshay	. . .	190
303 ,, ,, Wardlow	. . .	192
304 Arrow-head, bronze	. . .	192
305 Knife	. . .	192
306 ,, Wetton	. . .	193
307 Fibula, Waleby	. . .	194
308 ,, Elton	. . .	194
309 ,, Monsal Dale	. . .	194
310 to 315 Fibulæ, various localities	. . .	194
316 Fibula, Royal Irish Academy	. . .	194
317 ,, Little Chester	. . .	194
318 ,, Bottisham	. . .	195
319 ,, Elton	. . .	195
320 ,, Hoylake	. . .	195
321 Armillæ, Stony Middleton	. . .	196
322 Torques, Needwood Forest	. . .	198
323 ,, Royal Irish Academy	. . .	199
324 Horse-shoe, lamp, and fibulæ, Gloucester	. . .	201

ANGLO-SAXON PERIOD.

325 Plan of interment, Lapwing Dale	. .	209
326 Pottery, Cinerary Urns, Kingston	. .	215
327 ,, ,, North Elmham	. .	217
328 ,, ,, Chestersovers	. .	217
329 ,, drinking vessel	. .	217
330 ,, Cinerary Urns, Selzen	. .	221
331 ,, ,, Londinières	. .	221
332 ,, ,, Selzen	. .	221
333 ,, ,, Londinières	. .	221

LIST OF ILLUSTRATIONS.

Fig.		Page
334	Pottery, Cinerary Urn, Cologne	221
335	,, ,, Pfahlbau	223
336	,, ,, ,,	223
337	,, ,, ,,	223
338	,, ,, ,,	223
339	,, ,, Kings Newton	222
340	,, ,, ,,	222
341	,, ,, ,,	224
342	,, ,, ,,	224
343	,, ,, ,,	224
344	,, ,, ,,	225
345	,, ,, ,,	225
346	,, ,, ,,	225
347	,, ,, ,,	226
348	,, ,, ,,	226
349	,, notched stick	227
350	,, ,,	227
351	,, punctured ornament, Kings Newton	227
352	,, cup, Kings Newton	227
353, 354, 355	Glass tumblers	229
356, 357	Ale-glasses	229
358	Glass cups	229
359	Cup-bearer, with ale-glass	230
360	Cellarer, with barrels and pitchers	230
361	Banquet	231
362, 363	Glasses, decanter shape	231
364, 365, 366	Ale-glasses	232
367, 368, 369	Glass and clay beads	233
370	Glass and amber necklace	234
371	,, Bead and ring	235
372	Sword, Tissington	237
373	,, Grimthorpe	237
374	,, ,, guard	237
375	,, ,, chape	237
376 to 390	Swords, from illuminated MSS.	239
391	Swordsman with shield and sword	240
392 to 396	Knives or daggers, Kentish graves.	242
397 to 403	Spear-heads, Kentish graves, etc.	244
404	Spears, from illuminated MSS.	244
405	Shield plates, etc., Grimthorpe	246
406	Umbone of shield, Kentish graves	247
407	,, ,,	247

LIST OF ILLUSTRATIONS.

Fig.
408	Umbone of shield, Tissington	247
409 to 416	Shields, from illuminated MSS.	248
417 to 428	Buckles, from Kentish graves	249, 250
429	Drinking-cup, Benty Grange	251
430	Enamelled ornaments, etc., Benty Grange	251
431	Helmet, Benty Grange	253
432	Ornaments, Benty Grange	253
433	Plan of interment, Barlaston	259
434	Bronze ring, Barlaston	259
435	Enamelled disc, Barlaston	260
436	,, ,, Middleton Moor	261
437	,, ornament, Middleton Moor	262
438	,, ,, Royal Irish Academy	262
439	Bronze disc and rivets, Grimthorpe	263
440	Horse-shoe, Berkshire	264
441	,, ,,	264
442	Plan of interment, Rhine	265
443	Fibula, Kingston Down	267
444	,, Winster Moor	269
445	Pendant cross, Winster Moor	269
446	Fibula, Sittingbourne	270
447	,, Wingham	270
448	,, Kent	270
449	,, Stowe Heath	271
450	,, Ingarsby	271
451	,, Northamptonshire	272
452	,, Stow Heath	272
453	,, Royal Irish Academy	272
454	,, ,,	273
455	,, ,,	273
456	,, Westmoreland	274
457	,, Bonsall	275
458	,, ,,	276
459	,, Westmoreland	277
460	Bucket, Northamptonshire	281
461	,, Fairford	281
462	,, Envermeu	282
463	Drinking-cup, Sibertswold	283
464	,, ,,	283
465	Bronze bowl, Over-Haddon	284
466	,, box, etc., Church Sterndale	285
467	,, ornament, ,,	285

Fig.		Page
468 Bronze box, Newhaven		286
469 Comb, Royal Irish Academy		287
470 ,, Kent		287
471 ,, Thames		288
472 ,, Arica		288
473 ,, ,,		288
474 ,, Indian scalp		288
475 ,, Tweezers, Leicestershire		289
476 Châtelaines, Kent		290
477 Latch-keys (?) Ozengall		291
478 Girdle suspenders, Wilbraham		291
479 ,, Searby		291
480 Hair-pin, Searby		292
481 ,, Royal Irish Academy		292
482 ,, ,,		292
483 ,, ,,		292
484 Draughtmen, Cold Eaton		294
485 Dice, Gilton		295
486 Quern, Winster		295
487 ,, Kings Newton		296
488 Triturating stones		296
489 ,,		296

INTRODUCTION.

THE object of the following work is, I apprehend, so obvious as to render an introduction scarcely needful. It may be well, however, to remark, that it is the only work of its kind which has ever been issued, and that therefore, taking a stand of its own, and following no other either in plan or treatment of its subject, it is hoped that it will command the attention of antiquaries and of all who are interested in the history and the manners and habits of our early forefathers.

It has long appeared to me that a general *résumé* of the almost endless store of knowledge presented by the very varied relics of the grave-mounds of the three great divisions of our history—the Celtic, the Romano-British, and the Anglo-Saxon—kept distinct from the histories of those peoples, and from extraneous matters, and treating them more in a general than in an ethnological manner, could not fail to be a useful addition to our archæological literature, and would prove of great value and convenience to the general reader, as well as to the antiquary and the historian. Thus it is that I have been induced to prepare the present volume.

I have treated my subject in a popular manner, divesting it of technicalities, of theories, and of discursive matter, and have endeavoured, as far as space would permit, to give, simply and clearly, as correct an insight as possible into the modes of burial adopted in early times in our own country, and into the various remains of different races which an examination of their grave-mounds discloses.

Having great faith in the usefulness of engravings, and believing that, if judiciously introduced, a work of the kind cannot be too profusely illustrated, I have brought together in my present volume a larger number of engravings than could well have been expected; and these, I trust, will add much to its usefulness and value. To all my kind friends who have aided me in this matter I give my hearty thanks.

The work may have some, perhaps many, defects. If such exist, I shall be thankful to have them pointed out, and to remedy them in a future edition.

LLEWELLYNN JEWITT.

Winster Hall,
    Derbyshire.

Grave-mounds and their Contents.

CHAPTER I.

Grave-mounds in General—Their Historical Importance—General Situation—Known as Barrows, Houes, Tumps, and Lows—List of Names—Division into Periods.

TO the grave-mounds of the early inhabitants of our island, more than to any other source, we are indebted for our knowledge of their arts, their habits, and their occupations. Indeed, to these mounds and their contents, we owe almost all the knowledge we possess as to the history of the races and peoples who have preceded us, and are enabled to determine, approximately, their chronological succession as masters of the soil.

From the very earliest ages men of every race have bestowed peculiar care over the graves of the dead, and have marked to later ages, in an unmistakable manner, these places of sepulture, which have, in many instances, been preserved with religious care to modern times. Thus the relics which they contain have come down to us intact, and even now tell their wondrous tale, in a language of their own, of ages and of races of beings long since passed away. A single implement of stone or of flint; a weapon or an ornament of bronze, of iron, or of bone; a bead of jet or of glass; an urn, or even a fragment of pottery; or any one of the infinity of other relics which are exhumed, no matter to what period they belong, or from what locality they may have come; one and all tell their own tale, and supply new links to our ever-extending chain of knowledge.

To the graves, then, of our earliest ancestors, must we mainly turn for a knowledge of their history and of their modes of life; and a careful examination and comparison of their contents will enable us to arrive at certain data on which, not only to found theories, but to build up undying and faultless historical structures.

As, wherever the country was populated, interments of the dead must, as a necessity, have taken place, these all-important store-houses of after-knowledge exist, or have existed, to more or less extent in almost every district throughout the land, and give evidence, whenever opened by experienced hands, of their historical value and importance. The earliest grave-mounds are mostly found in the mountainous districts of the land—among the hills and fastnesses; the latter overspreading hill and valley and plain alike. Thus, in Cornwall and Yorkshire, in Derbyshire and in Dorsetshire, in Wiltshire and in many other districts, the earliest interments are, or have been, abundant; while the later ones, besides being mixed up with them in the districts named, are spread over every other county. In the counties just named Celtic remains abound more than those of any other period. In Dorsetshire, for instance, that county, as the venerable Stukeley declares, "for sight of barrows not to be equalled in the world," the early mounds abound on the downs and on the lofty Ridgeway, an immense range of hills of some forty miles in extent, while those of a later period lie in other parts of the county. In Yorkshire, again, they abound chiefly in the wolds; and in Cornwall, on the high lands. The same, again, of Derbyshire, where they lie for the most part scattered over the wild, mountainous, and beautiful district known as the High Peak—a district occupying nearly one half of the county, and containing within its limits many towns, villages, and other places of extreme interest. In this it resembles Dorsetshire; for in the district comprised in the Ridgeway and the downs are very many highly interesting and important places, around which the tumuli are most plentiful.

It is true that here and there in Derbyshire, as in other counties, an early grave-mound exists in the southern or lowland portion of the county; but, as a rule, they may be almost said to be peculiar, and confined, to the northern, or hilly district, where in some parts they are very abundant. Indeed, there are districts where there is scarcely a hill, even in that land, where

> "Hills upon hills,
> Mountains on mountains rise,"

where a barrow does not exist or is not known to have existed. In passing along the old high-road, for instance, over Middleton Moor by way of Arbor Low,* Parcelly Hay, High Needham, Earl Sterndale, and Brier Low, to Buxton, or along the high-roads by way of Winster, Hartington, or Newhaven, the practised eye has no difficulty in resting on the forms of grave-mounds on the summits of the different hills or mountains, whose outlines stand out clear and distinct against the sky.

The situations chosen by the early inhabitants for the burial of their dead were, in many instances, grand in the extreme. Formed on the tops of the highest hills, or on lower but equally imposing positions, the grave-mounds commanded a glorious prospect of hill and dale, wood and water, rock and meadow, of many miles in extent, and on every side stretching out as far as the eye could reach, while they themselves could be seen from afar off in every direction by the tribes who had raised them, while engaged either

---

* Of this stone circle, one of the next in importance to Stonehenge, an account will be given in a future chapter.

in hunting or in their other pursuits. They became, indeed, land-marks for the tribes, and were, there can be but little doubt, used by them as places of assembling.

Sepulchral tumuli are known as barrows, lows, houes, tumps, etc. *Barrow* is of pretty general use; *low* is almost universal in Derbyshire, Staffordshire, and other districts; *tump* is in use in Gloucestershire, etc.; and *houe* in Yorkshire.

In Derbyshire and Staffordshire, the term "Low" is so very usual that, wherever met with, it may be taken as a sure indication of a barrow now existing or having once existed at the spot. As a proof of this, it will only be necessary to say that at about two hundred places in Derbyshire alone, and at about half that number on the neighbouring borders of Staffordshire, which bear the affix of *Low*, barrows are known to exist or have already been opened. For my present purpose, it will be sufficient to give the few following names:—Arbor-Low, Kens-Low, Ringham-Low, Blake-Low, Fox-Low, Gib-Low, Green-Low, Great-Low, Grind-Low, Cal-Low, Chelmorton-Low, Casking-Low, Larks-Low, Thirkel-Low, Ribden-Low, Har-Low, Bas-Low, High-Low, Foo-Low, Lean-Low, Huck-Low, Borther-Low, Dow-Low, Totman's-Low, Staden-Low, Stan-Low, Blind-Low, Boar-Low, Bottles-Low, Brown-Low, Caldon-Low, Calver-Low, Cock-Low, Cop-Low, Cow-Low, Cronkstone-Low, Dars-Low, Drake-Low, Elk-Low, End-Low, Far-Low, Pike-Low, Fowse-Low, Galley-Low, Gris-Low, Grub-Low, Herns-Low, Hawks-Low, Horning-Low, Hard-Low, Knock-Low, Knot-Low, Laidmans-Low, Lady-Low, Liffs-Low, Lomber-Low, Lousy-Low, Mick-Low, Moot-Low, Money-Low, Musden-Low, May-Low, Needham-Low, Nether-Low, Ox-Low, Off-Low, Pars-Low, Painstor-Low, Peg-Low, Pigtor-Low, Pike-Low, Pinch-Low, Queen-Low, Ravens-Low, Rains-Low, Rick-Low, Rocky-Low, Rolley-Low, Round-Low, Rusden-Low, Saint-Low, Sitting-Low, Sliper-Low, Thoo-Low Three-

Lows, Ward-Low, Warry-Low, White-Low, Whithery-Low, Wool-Low, and Yarns-Low. To some of these I shall again have occasion to make reference. In Yorkshire, the names of William Houe, Three Houes, and Three Tremblers Houes, will be sufficient indication of the local use of the term "Houe."

Grave-mounds may, naturally, be divided into the three great periods; the Celtic, the Romano-British, and the Anglo-Saxon. This division will be adopted in the present volume, and it will be its aim, while speaking of the characteristics of each, to classify and describe their contents, and to point out, briefly, such circumstances of interment, and such evidences of customs, as they may present, and which may appear to be of sufficient interest and importance to its plan.

Of the forms of barrows, and their characteristics and modes of construction, occasion will be taken to speak in a later chapter.

CHAPTER II.

Ancient British, or Celtic, Period—General characteristics of the Barrows—Modes of construction—Interments by inhumation and by cremation—Positions of the body—Hitter Hill Barrow—Elliptical Barrow at Swinscoe—Burial in contracted position—In sitting and kneeling positions—Double interments.

THE barrows of the Celtic, or ancient British, period vary in their form and size as much as they do in their modes of construction, and in their contents. Sometimes they are simply mounds of earth raised over the interment; sometimes heaps of stones piled up over the body; and sometimes again a combination of cist and earth and stone. Generally speaking the mounds are circular, rising gradually and gently from the level of the ground towards the centre, but in some instances the rise is somewhat acute. Now and then they are oval in form. Where elliptical barrows occur (generally known as "long barrows"), they are, I have reason to believe, not matters of original design, but of accident, through additional interments; and I much doubt the propriety of archæologists at the present day continuing the very questionable nomenclature adopted by Sir R. C. Hoare and others. In some cases, however, as in the instances of chambered or walled tumuli, the elliptical form of the barrow can be easily understood. An examination of a very large number of barrows leads me to the opinion that the original form of all was circular, and that no deviation from that form, and no difference in section, can be taken as indicative of period or of race.

The other appellation occasionally used, of "twin barrows," is further evidence of this—two interments having been made within a short distance of each other, and the mounds raised over them running into and joining each other. It may, however, for purposes of description, and for this alone, be well to retain the names, while discarding much of the theory and of the system which has been attempted to be established regarding them.

The mounds of earth alluded to, present occasionally highly interesting and curious features, and show that, like those of a different construction, they have frequently been used for successive interments. The section of one of these is shown on the next page. It is one of a group of six barrows on Lord's Down, in the parish of Dewlish, in Dorsetshire. It was eighty-two feet in diameter, and fourteen feet in height in the centre. The primary interment, an urn, was placed in a cist cut in the chalk sub-soil. Over the urn was raised a small cairn of flints, and the cist was then filled in, and raised a little above the surface with chalk rubble. Over this was a layer of earth, upon which an interment had taken place, and in its turn covered with a thick layer of chalk rubble, in the centre of which, in a cist, another interment had again been made. Above this rose another layer of earth, another of chalk, and then a final one of earth, on each of which interments had at different periods been made. Thus the tumulus, which was formed of alternate layers of chalk and earth, exhibited no less than six successive sepulchral deposits.* The interments were both by inhumation and cremation.

Another example of a barrow of this period is shown in section on the fig. 3. There had originally been four

* This remarkable barrow was excavated by Mr. Warne, and a fully detailed account given by him in his valuable work, the "Celtic Antiquities of Dorset," from which the illustration is taken.

8 GRAVE-MOUNDS AND THEIR CONTENTS.

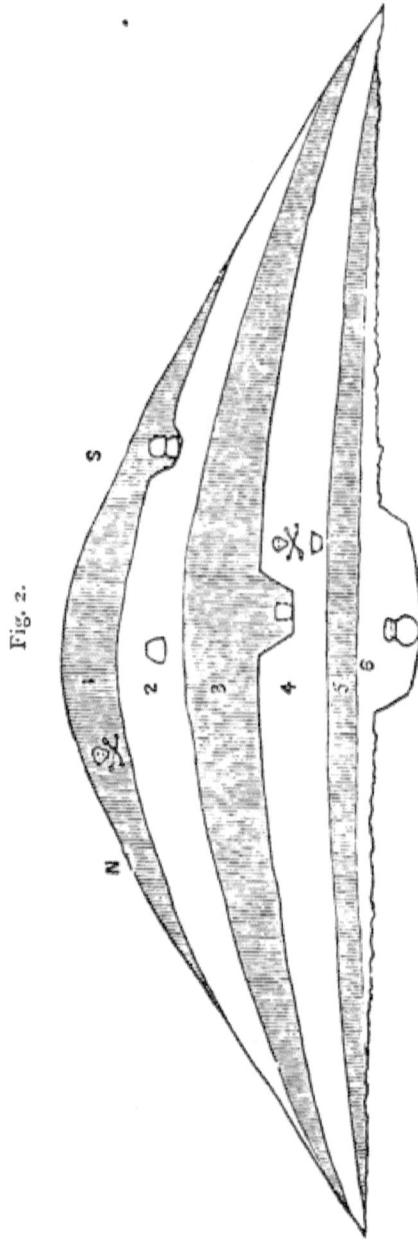

Fig. 2.

1. Earth to the depth of three feet.
2. Chalk, two feet in depth.
3. Earth to the depth of three feet.
4. Loose rubble chalk, three feet in depth.
5. Earth, one foot in depth
6. Chalk rubble, six inches in depth.
Cist, filled with chalk rubble and flints, two feet in depth.

small mounds, or barrows, formed in a group, on the natural surface of the ground (see the two dotted lines in the engraving). They were composed of tempered earth, approaching in tenacity almost to clay, and on these the general mound was raised to a height of about eighteen

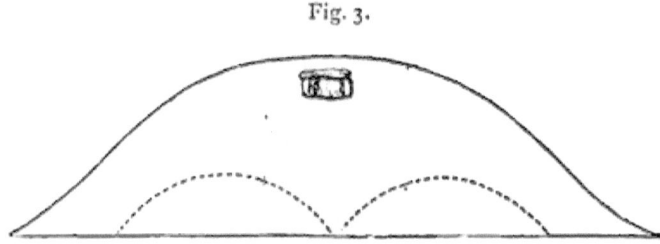

Fig. 3.

feet, and was composed of earth, intermixed with loose rubbly limestones. Nearly at the top, in the centre, a stone cist, enclosing an interment, was discovered. It was in form a perfect, though miniature, example of what are commonly called cromlechs.

It is not an unfrequent occurrence in barrows to find that interments have been made at different periods and by different races, as will be hereafter shown.

The two examples of mounds of earth already given will show the successive layers which have occurred in their formation. The simpler, and intact, mounds of earth, which are very common, require no illustration. They are simply immense circular heaps of earth raised over the interment, whether in cist or not.

Barrows, or mounds, of stone are of frequent occurrence. They are of very simple construction. The interment, whether by cremation or otherwise, having been made in a natural or artificial cist, or simply laid upon the natural surface of the ground, rough stones were placed in a large circle around it, and an immense quantity of stones were then piled up to a height of several feet. Some of these cairns are of very great size, and cover a large area of ground. Sections of two tumuli of this description are given in figs. 4 and 5. The cairn of stone was, as will be

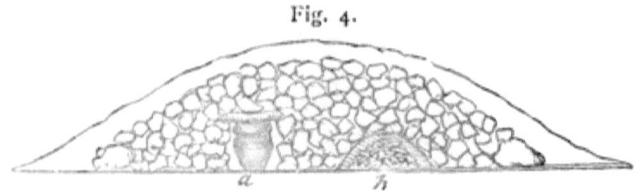

Fig. 4.

seen, covered to some depth with earth; perhaps in some

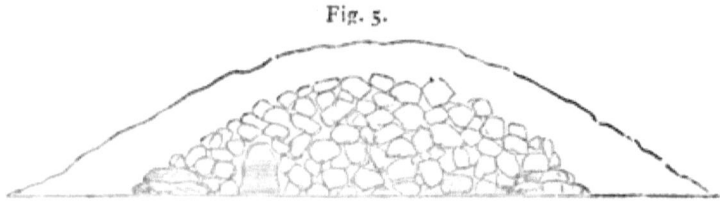

Fig. 5.

instances this might be a part of the original design, but in most cases the soil which now covers these stone barrows may be traced to the ordinary process of decay of vegetation in successive ages.

Barrows were not unfrequently surrounded by a circle of stones, set upright in the ground. These circles, in many instances, remain to the present day in different parts of the kingdom, and, the barrow itself having disappeared, are commonly called by the general appellation of "Druidical circles." But of these, later on. The construction of the stone circles varied considerably. In some instances the upright stones were pretty close together; in others, wide apart; and in others, again, the spaces between the uprights were filled in with a rude loose rubble masonry, which thus formed a continuous wall.

Some tumuli contained stone chambers and passages, formed of massive upright slabs, and covered with immense blocks of stone. Over these chambers, etc., the mounds of earth, or of stone, or of both combined, were raised, as will be hereafter shown.

Interments in the Celtic grave-mounds were both by inhumation and cremation, and the modes of interment, in both these divisions, was very varied.

Where inhumation obtained, the body was sometimes laid on its side, in a contracted position; at others, extended full length on its back or side; and in other instances, again, was placed in an upright sitting or kneeling posture. Occasionally, too, where more than one body has been buried at the same time, they have been laid face to face, with their arms encircling each other; at other times an infant has been placed in its mother's arms.

When cremation has been practised, the remains have either been gathered together in a small heap on the surface of the ground—sometimes enclosed in a small cist, at others left uncovered, and at others covered with a small slab of stone—or wrapped in a cloth or skin (the bone or bronze pin which has fastened the napkin being occasionally found), or enclosed in cinerary urns, inverted or otherwise. In some instances, even when placed in urns, they were first enclosed in a cloth.

These are the general characteristics of the interments of the Celtic period, and they will be best understood by the following examples.

When the body has been buried in a contracted position, it is found lying on its side; the left side being the most usual. The head generally inclines a little forward; the knees are drawn up near to the chest, and the heels to the thighs; the elbows are brought near to the knees,—frequently, indeed, one of them will be found beneath, and the other on, the knees, which have thus been held between them; and the hands are frequently brought up to the front of the face. This position, which is after all, perhaps, the most easy and natural one to choose, will be best understood by the following engraving (fig. 6), which shows an interment found in a barrow on Smerril Moor, opened by my much lamented friend the late Mr. Thomas

Fig. 6.

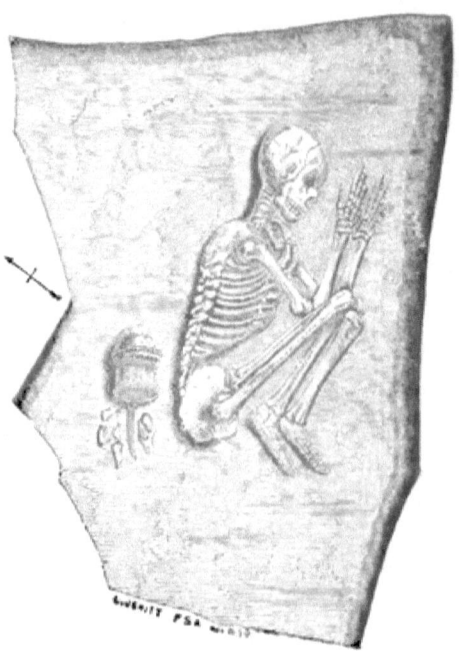

Bateman. In this case the body had been laid on its left side in an irregularly formed cavity on the surface of the natural rock, on a bed of clay, over which, as usual, the mound was formed of loose stones and mould. Behind the skeleton, as will be seen in the engraving, was found a remarkably fine "drinking cup," along with a bone meshing rule or modelling tool, twelve inches long, made from the rib of a horse or cow; a flint dagger; an arrow-head; and some other implements, also of flint, all of which had been burned. The femur of this skeleton measured nineteen and a half inches, and the tibia sixteen inches.

The next example (fig. 7), from Tissington, will be seen to have been laid in very much the same position.

Fig. 7.

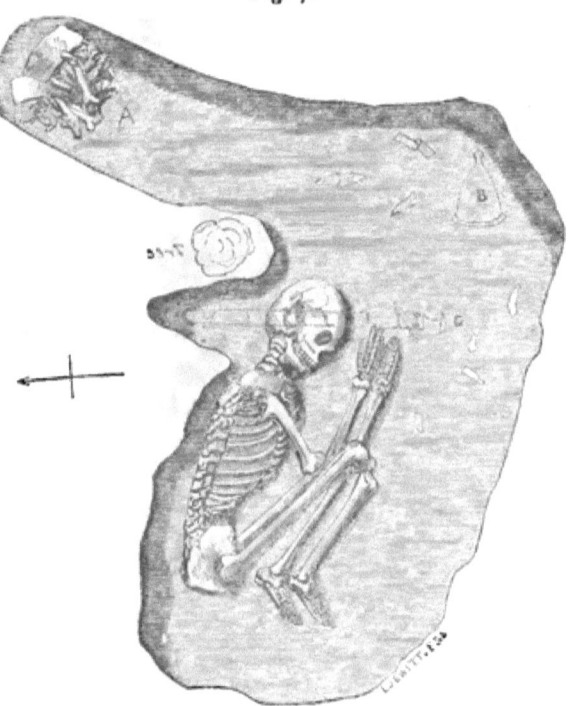

It lies on its left side, the knees drawn up, and the feet, elbows, and hands in the position I have already described. This barrow possessed considerable interest, from the fact that a later interment—of the Anglo-Saxon period—had been made immediately above the figure here subjoined. To this I shall refer under the head of Anglo-Saxon. In the same barrow, shown at A on the engraving, was a deposit of burnt bones. Fig. 8, again, shows an interment in the contracted position, the head, in this instance, resting upon the left hand. The skeleton lay in an oblong

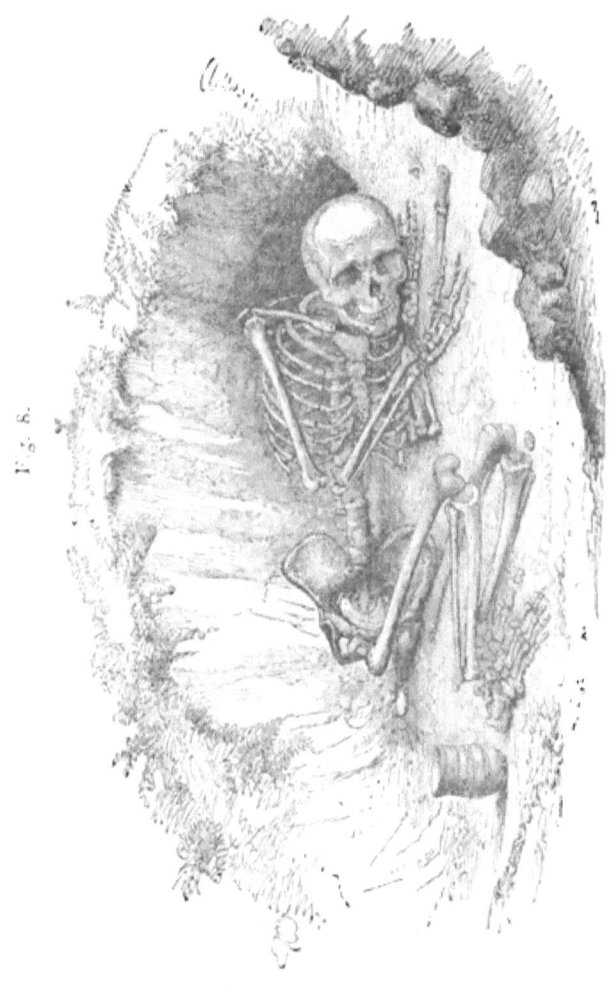

oval cist, five feet long by two and a half feet wide, smoothly hollowed out of the chalk, and over this the mound was raised.

POSITIONS OF THE BODY. 15

Fig. 9.

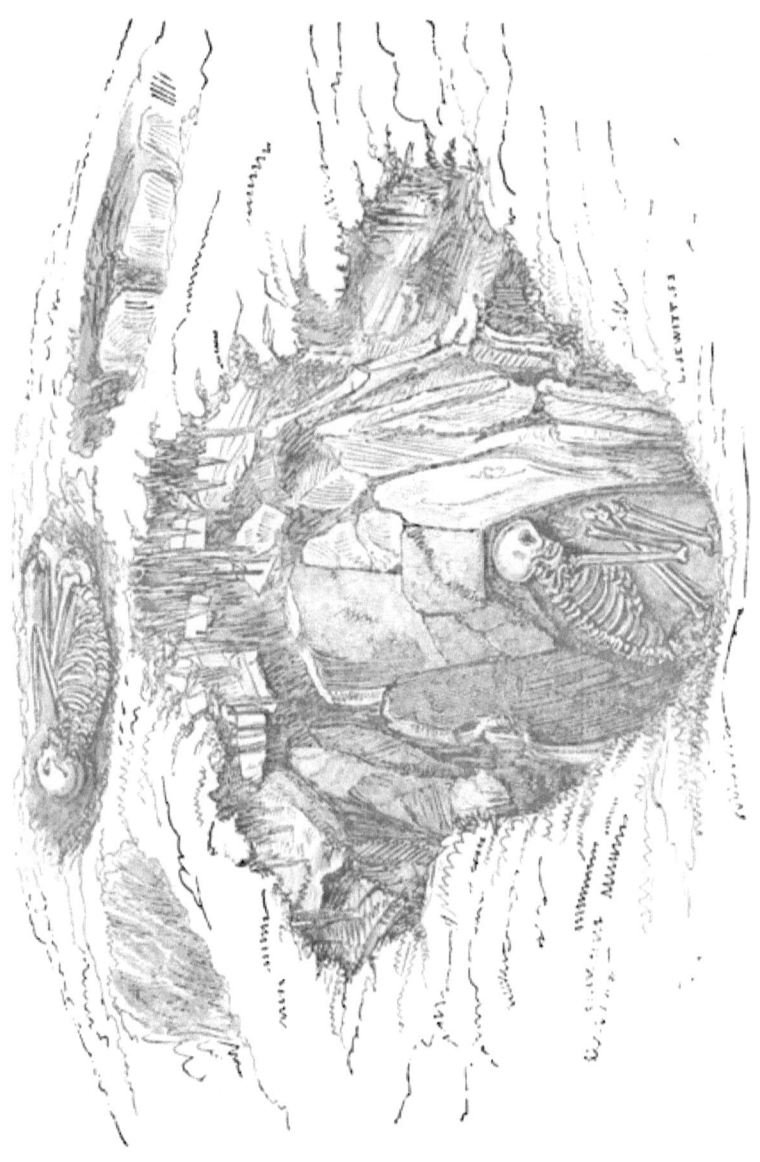

Along with it were found a bronze dagger, a barbed arrow-head of flint, a beautiful drinking cup and other objects. This example is from Roundway Hill, in North Wiltshire.* Another excellent example, from Hitter Hill, Derbyshire, is given in the next engraving (fig. 9), which shows successive interments, each being on the left side, in the usual contracted position.

Of this barrow, the opening of which presented peculiarly interesting features, a tolerably detailed account will be advisable. It was opened by myself and Mr. Lucas in 1862. The mound, which was about twenty-two feet in diameter, was composed of rough stone and earth intermixed. It was only about three feet in height, its centre being somewhat sunk. The first opening was made at the part marked A on the accompanying ground-plan (fig. 10), where, from the outside, we cut a trench, four feet in width, in a north-easterly direction, towards the centre of the barrow, and soon came upon an interment of burnt and unburnt human bones. Along with these were an immense quantity of rats' bones† and snail-shells. After proceeding to a dis-

---

* See Crania Britannica, one of the most valuable ethnological works ever issued.

† It will be well to bear in mind that when " rats' bones" are mentioned, it must be understood that they are the bones, not of the common rat, but of the water-vole or water-rat. They are very abundant in Derbyshire barrows, and, indeed, are so frequently found in them, that their presence in a mound is considered to be a certain indication of the presence of human remains. " The barrows of Derbyshire, a hilly, almost mountainous, county, abounding with beautiful brooks and rills, inhabited by the water-vole, were made use of for its *hybernacula*, or winter retreats, into which it stored its provisions, and where it passed its time during the cold and frosty season. It is a rodent, or gnawer, or vegetable eater, and, as I have described elsewhere, has a set of grinding-teeth of the utmost beauty, and fitted most admirably for the food on which it lives. The part of the matter which is curious to the antiquary is, that the bones in Derbyshire barrows are frequently perceived to have been gnawed by the scalpri-form incisors of these animals. I have endeavoured to explain, in the note referred to, that all the rodents amuse themselves, or possibly preserve their teeth in a naturally useful state, and themselves in health, by gnawing any object that comes in their way. This is well known to every boy who keeps rabbits. I remember, some

tance of seven feet, we came upon the side, or what may almost be called the entrance, of a cist formed partly of the

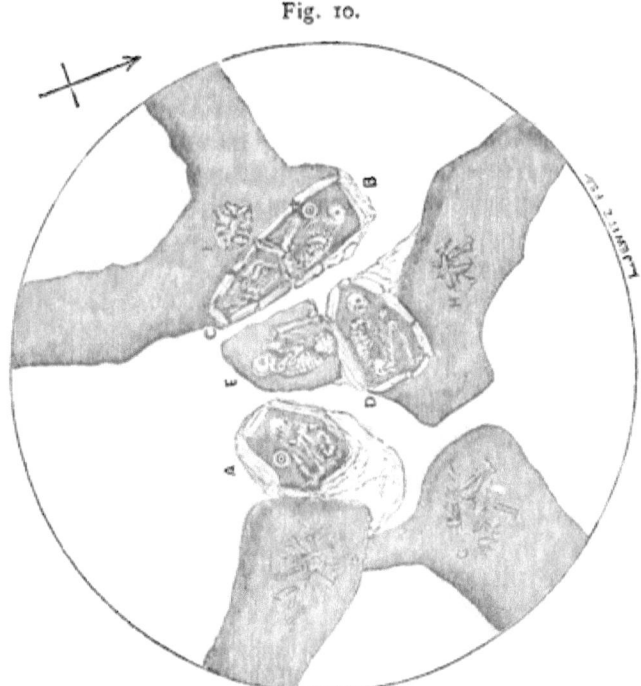

Fig. 10.

natural rock, and partly of stones set up edgewise. The dimensions of this cist were about forty inches by twenty-

years ago, seeing a very fine black squirrel in the house of a workman in this town, which had been sent him by his son from Canada. It was found that it was impossible to keep this animal in any wooden house. He would gnaw a road out of the strongest wooden cage that could be made for him, in a few hours. In consequence, his owner made him a *tin cage*, in which he was kept securely. In confirmation of what I have said respecting the water-voles, vegetable feeders, gnawing the bones of the ancient Britons in barrows, I may refer to Linnæus's most interesting *Tour in Lapland.* When in Lycksele, Lapland, June 1, he describes the *Kodda*, or hut of the Laplander, and incidentally remarks, " Everywhere around the huts I observed horns of the reindeer lying neglected, and it is remarkable that they were gnawed, and sometimes half devoured, by six inches, and it was two feet in depth, the floor being three feet six inches below the surface. The cist was formed between two portions of natural rock, and protected at its entrance by a large flat stone set up edgwise, and other stones filled up the interstices at the sides. It was also covered with a large flat stone. On clearing away the surrounding earth, after removing the covering stone, we were rewarded by finding that the cist contained the fragmentary remains of a young person, which had lain on its right side, in the usual position, with the knees drawn up.

Fig. 11.

The accompanying engraving (fig. 11) will show the opened cist, with the stone across its entrance, and the interment *in situ*. In front of the skeleton, and close to its hands, was a remarkably good and perfect food vessel, which was richly ornamented with the diagonal and herring-bone lines, formed by twisted thongs impressed into the soft clay.

squirrels."—I. 127. That is, if anything were truly devoured, it was the antlers, not the bodies. "The bones of the *Arvicola*, or water-vole, were found in the exploration of the colossal tumulus of Fontenay de Marmion, which was one of the galleried tumuli, opened in 1829, near Caen in Normandy. It belonged to the primeval period of the ancient Gauls.— Mem. de la Soc. des Antiq. de Normandie, 1831-3, p. 282."—*Dr. Davis.*

The next morning we dug a trench four feet wide, on the west side towards the centre, as shown at B on the plan (fig. 10), and the day's labours had an equally satisfactory result. At about the same distance as on the previous day we came to the side of a cist, immediately in front of which, at F on the plan, lay a heap of burnt bones, and a few flakes of burnt flint. Having cleared away the surrounding stones and earth, and removed the large flat covering stones, which showed above the surface of the mound, we found the cist to be composed on one side by the natural rock, and on the other by flat stones set up on edge. Its dimensions were about one foot ten inches by four feet, and it contained a large quantity of rats' bones and snails' shells. In this cist was an interment of an adult, much crushed by one of the large covering stones having fallen upon it. Thanks to this circumstance, however, a food vessel, which we discovered, owed its preservation. The body lay in the usual contracted position, on its right side, as shown on the ground-plan at B, and in front and close to the hands was the food vessel, which, like the other, was taken out entire. It is five and a quarter inches in height, and six and a quarter inches in diameter at the top, and is richly ornamented.

Continuing the excavations to the south, we found that another cist C adjoined the one just described, and was, like it, formed of flat stones set up edgewise; in fact, it was like one long cist divided across the middle. In this second cist, besides the usual accompaniment of rats' bones, was the remains of an interment, sufficiently *in situ* to show that the skeleton had, like the others, been deposited in a contracted position. A small fragment of pottery was also found, but owing to the cist being so near the surface the stones had been partially crushed in, and thus both the deposit and the urn had become destroyed. A portion of a stone hammer was also found.

The two cists are here shown (fig. 12), which also shows

the central interment at a higher level, to be hereafter described.

On the following Monday we resumed our operations by making an opening on the north-west side, as shown at D on the plan. Here, again, at a few feet from the outer edge, we came upon an interment H, without a cist, accompanied by an unusual quantity of rats' bones. Continuing the excavation, we were again rewarded by the discovery of a fine cist, but at a greater depth than those before described. Above this cist we found some large bones of the ox, and on the

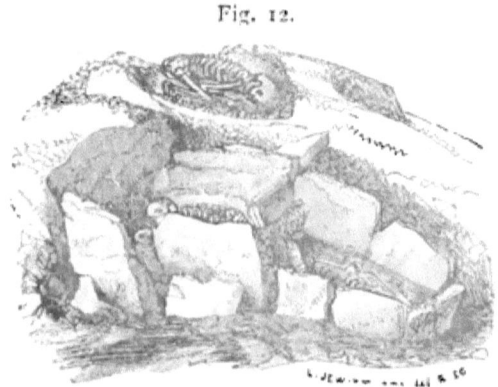

Fig. 12.

covering stone was a deposit of burnt bones and ashes, with innumerable quantities of rats' bones.

The cist, which was covered with one extremely large flat stone, we found to be formed partly of the natural rock, and partly—like the others—of flat stones set up edgewise; and it was, without exception, the most compact and neatly formed of any which have come under our observation. Its form will be seen on the plan at D, and its appearance, when the interior soil was removed, is shown on fig. 9. The dimensions of the cist were as follows:—Width at the foot, twenty-four inches; extreme length, forty inches; general depth, twenty inches. The floor was composed of the natural surface of the rock, with some small flat stones laid to make it level, and at the narrow end a raised edge of stone, rudely hollowed in the centre, formed a pillow on which the head rested. The sides of the cist were square on the one side to the length of twenty-eight, and on the other of twenty-one, inches, and it then gradually became narrower until at the head its width was only ten inches. When the cist was cleared of its accumulation of soil and

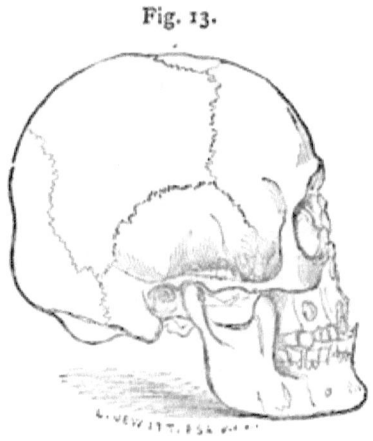

Fig. 13.

rats' bones—of which scores of jaw-bones were present, thus showing the large number of these ravaging animals which had taken up their abode there—it presented one of the most beautiful and interesting examples of primeval architecture ever exhumed. It contained the skeleton of an adult, laid on his left side, in the usual contracted position, but without any pottery or flint. The skull, of which an outline engraving is given on fig. 13, is a most interesting and characteristic example of the cranium of an ancient Coritanian Briton. It is brachy-cephalic, and is the subject of deformity from nursing on the cradle-board in infancy.* It is the skull of a middle-aged man, and is remarkably well formed. The bones, with the exception of some of the small ones, were all remaining, and formed a skeleton of considerable ethnological interest. The small bones were gnawed away by the rats, and it is curious to see to what distances, in some interments, these active little animals have dragged even large bones from their original resting-places. It may not be without interest to note, that within the skull of this skeleton the bones of a rat, head and all, were found imbedded in the soil, along with some small stones, which he doubtless had dragged in with him on his last excursion. We continued our excavations in a north-easterly direction, as shown at G on the plan, and found another interment, but without a cist or any other notable remains; and next day we commenced opening that portion of the centre of the barrow between the cists already described, and soon came upon an interment of an adult person, as shown on the plan at E. The bones were very much disturbed, but sufficient remained to show that the deceased had been placed on his left side, in the same contracted position as the others in this mound. The body was not more than twelve inches below the surface, and was much disturbed, but it is more than probable the top of the barrow had at some distant time been taken off, most likely for the sake of the stone. The position of this interment will be seen on reference to the plan, and it is also shown on figs. 9, 11, and 12.

In addition to these illustrations, it will be sufficient to give the annexed engraving (fig. 14), which shows the position of a number of interments uncovered by Mr. Bateman in the centre of an elliptical barrow † at Swinscoe. The

---

* See Note on the Distortions which present themselves in the crania of the Ancient Britons, by J. Barnard Davis, M.D., in the "Natural History Review" for July, 1862, page 290.

† The elliptical form was evidently, in this case, the result of accident.

interments were as follows:—1. A young adult, lying in a contracted position, on its right side, in a shallow grave cut about six inches deep in the chert rock, with a stone placed on edge at the head, and another at the feet. 2. A young adult, lying on its right side, an upright stone at its head. 3. A middle-aged person, lying with the face upwards, and guarded

Fig. 14.

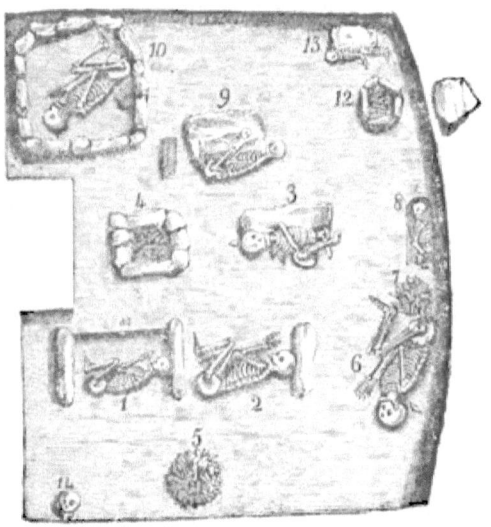

by a large stone at its side. 4. The bones of a young hog, enclosed in a stone cist. 5. Remains of a cinerary urn and burnt bones. 6. Skeleton of an aged man, lying in a contracted position, on its left side, upon a thin layer of charred wood. 7. A deposit of burnt bones. 8. Skeleton, very fragmentary. 9. A double interment, consisting of two skeletons, with a flat stone on edge by their side. These were an adult, and a child of a few months old only. 10. Skeleton of

The original mound had been circular, but the elongated form had been the consequence of successive interments.

an aged man, lying in the usual position, on his left side, enclosed in a circle of stones. Behind him lay a handsome drinking-cup. 12. Portions of a skeleton in a pentagonal cist. 13. Skeleton of a young person, placed close up to an upright flat stone. 14. Skull and portions of a skeleton. Several flints and other remains were found with these various interments.

Interments where the body has lain extended are of much rarer occurrence, in this period, than those just described. Some few instances have been brought to light in Derbyshire, and in other counties, but they are indeed " few and far between," and are the very rare exceptions to a general rule. One of the most interesting of these instances is the one called " Shuttlestone Low," opened a few years ago by Mr. Bateman, and thus described by him : " It consisted of a compact mass of tempered earth down to the natural surface of the land, below which point, in the centre of the barrow, there appeared a large collection of immense limestones, the two uppermost being placed on edge, and all below being laid flat, though without any other order or design than was sufficient to prevent the lowest course resting upon the floor of the grave inside which they were piled up, and which was cut out to the depth of at least eight feet below the natural surface; thus rendering the total depth, from the top of the mound to the floor of the grave, not less than twelve feet. Underneath the large stones lay the skeleton of a man, in the prime of life, and of fine proportions, apparently the sole occupant of the mound; who had been interred while enveloped in a skin of dark-red colour, the hairy surface of which had left many traces both upon the surrounding earth and upon the verdigris, or patina, coating; a bronze axe-shaped celt and dagger deposited with the skeleton. On the former weapon there are also beautifully distinct impressions of fern leaves, handfuls of which, in a compressed and half-decayed state, surrounded the bones from head to foot. From these leaves being discernible on one side of the celt only, whilst the other side presents traces of leather alone, it is certain that the leaves were placed first as a couch for the reception of the corpse, with its accompaniments, and after these had been deposited, were then further added in quantity sufficient to protect the body from the earth. The position of the weapons with respect to the body is well ascertained, and is further evidenced by the bronze having imparted a vivid tinge of green to the bones where in contact with them. Close to the head were one small black bead of jet and a circular flint; in contact with the left upper arm lay a bronze dagger with a very sharp edge, having two rivets for the attachment of the handle, which was of horn, the impression of the grain of that substance being quite distinct around the studs. About the middle of the left thigh-bone was placed the bronze celt which is of the plainest axe-shaped type. The cutting edge was turned towards the upper part of the person, and the instrument itself has been inserted vertically into a wooden handle, by being driven in for about two inches at the narrow end—at least the grain of the wood runs in the same direction as the longest dimension of the celt,—a fact not unworthy of the notice of any inclined to explain the precise manner of mounting these curious implements. The skull—which is decayed on the left side, from the body having lain with that side down—is of the platy-cephalic form, with prominent parietal tubers—the femur measures $18\frac{1}{2}$ inches."

Another good instance is from Yorkshire, where two skeletons were found side by side, extended, with their heads respectively east and west, lying in a bed of charcoal.

Occasionally, as has been stated on a previous page, interments have been made in an upright sitting position. Instances of this kind are rare and very curious. Our engraving (fig. 15) shows an interment of this kind, which occurred in a barrow at Parcelly Hay. The body had been placed in a sitting posture, leaning back against the side of the cist, which was only three feet in height, and not more than that in its greatest width. The cist was roughly covered in with large slabs of limestone. The skull, which was a remarkably fine one, has been engraved in "Crania Britannica."* On the covering-stones lay another skeleton, among

Fig. 15.

the loose stones of which the barrow itself was composed. This secondary interment was accompanied by a fine axe-head of stone and a bronze dagger. Of course the seated skeleton must have been of an earlier date still.

* Plate II., Decade 1.

Another remarkable instance of this kind of interment—but this time in a kneeling position—was discovered in the Cromlech De Tus, or De Hus, in Guernsey, by Mr. F. C. Lukis.* This interesting relic is situated near Paradis, in the parish of Vale, and is a chambered tumulus of simple but excellent construction. The mound is surrounded by a circle of stones, about twenty yards in diameter. In the centre is the principal chamber, covered with large flat stones, and from it to the extremity of the mound, on the east, was a passage formed by upright stones, and covered here and there with cap-stones dividing it into chambers. On the north side of this passage was a chamber formed by upright stones, on which rested the large flat covering stone; and close to this was another similar but smaller chamber. In the first were discovered "vases, bone instruments, celts, and human remains." In the latter, on removing the soil at the top, "the upper part of two human skulls were exposed to view. One was facing the north, and the other the south, but both disposed in a line from east to west,"—in other words, side by side, and shoulder to shoulder, but facing opposite ways. They were skeletons of adult males, and, on clearing away the soil, they were found to have been buried at the same time. "The perfect regular position of a person kneeling on the floor, in an upright posture, with the arms following the direction of the column, pelvis, and thigh-bones, and gradually surrounded by the earth, in like manner as may be conceived would be done were the persons buried alive, will give an exact representation of this singular discovery."

Another excellent example of this very unusual mode of interment—this time in a sitting posture—was discovered by some tufa-getters, and examined by Mr. Bateman, in

---

* Journal of the British Archæological Association, vol. i, p. 25.

Monsal Dale, and is shown on the accompanying engraving, fig. 16, which exhibits a section of the rock, etc.; and shows the position of the skeleton, and the manner in which the cavity containing the body had been filled up with the river sand. The body in this case, as in the last, had been placed in the cavity, in a sitting position, and must have been so placed from an opening in front. The cavity was ten or twelve feet above the bed of the river Wye, and above it were some five feet in thickness of solid

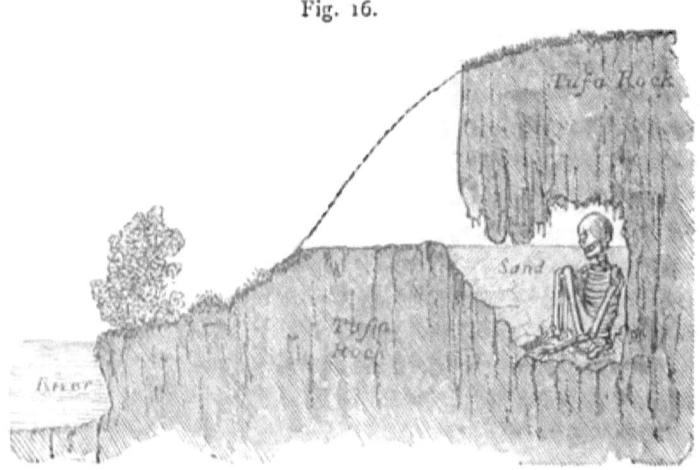

Fig. 16.

tufa rock, while, from the face of the rock, the cavity was about twelve feet. The body may therefore be said to have been entombed in the middle of the solid rock. The roof of the cavity when found was beautifully covered with stalactites. The skeleton was that of a young person, and near it were found a flint and some other matters. The cavity was filled to part way up the skull with sand.

Another example of interment in a sitting posture was discovered some years ago, at Kells in Ireland. These will be sufficient to show the curious character of this mode of interment.

DOUBLE INTERMENTS.

Several examples of double interments, besides those described above, have been discovered in different localities. One of the most curious is the one in the largest cist, in fig. 17. In this cist, which is composed of four upright

Fig. 17.

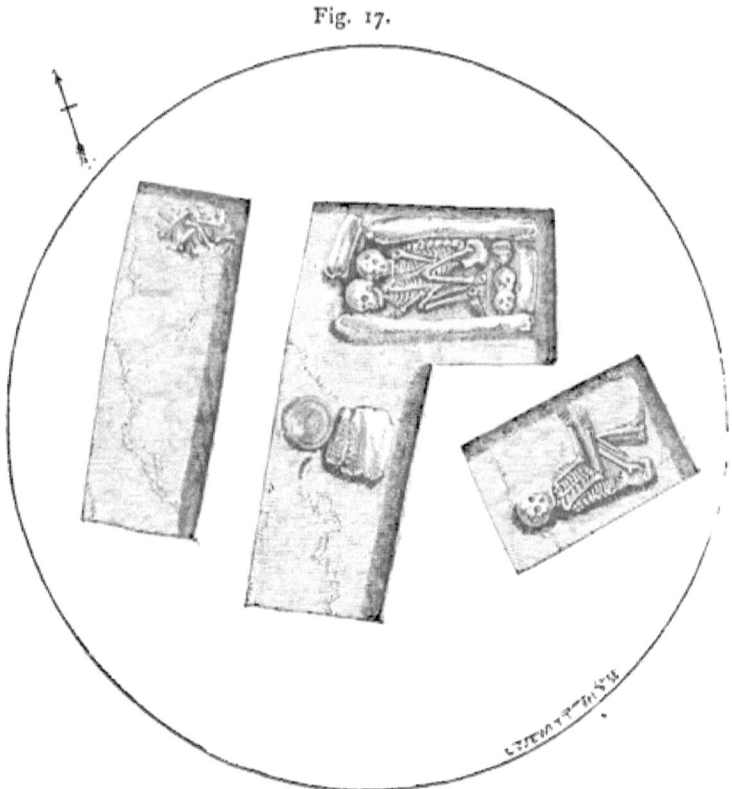

slabs of stone,* were the skeletons of a man and woman, and the remains of two children; the family having probably been immolated at the death of its head, and all buried together. A small urn was found in the same cist;

* "Ten Years' Diggings." p. 78.

and in the same barrow, in other portions which were excavated, as shown in the plan, were other interments, both by cremation and by inhumation.

In No. 9, on fig. 14, an interment of a mother and her child together is shown. Another instance is shown on the next engraving, fig. 18. In this instance the woman

Fig. 18.

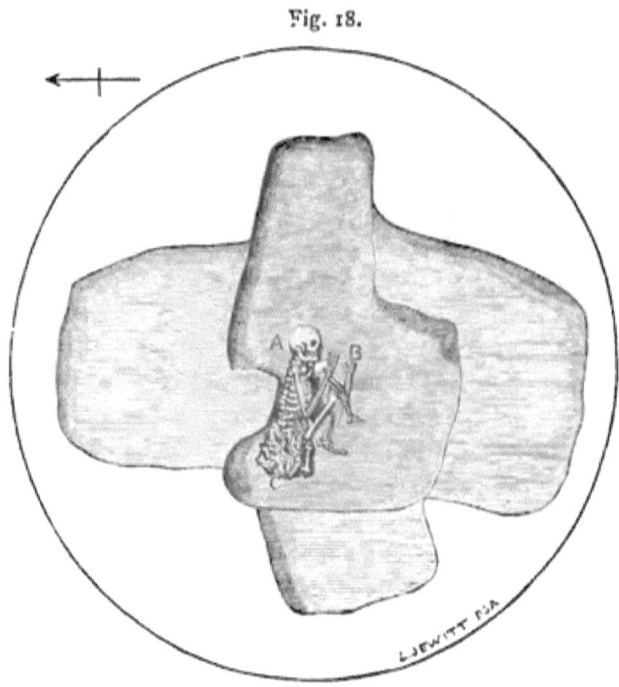

was laid in the usual contracted position, on her left side, with her head to the east. Close in front of the breast, lying in the arms in the same contracted position, lay the infant. Some flints and a fragment of pottery were found along with this touching interment.

CHAPTER III.

Ancient British or Celtic Period—Interment by cremation—Discovery of lead—Burial in urns—Positions of urns—Heaps of burnt bones—Burnt bones enclosed in cloth and skins—Stone cists—Long-Low—Liff's-Low, etc.—Pit interments—Tree-coffins.

WHEN the interment has been by CREMATION, the remains of the burnt bones, etc., have been collected together and placed either in a small heap, or enclosed in a skin or cloth, or placed in a cinerary urn, which is sometimes found in an upright position, its mouth covered with a flat stone, and at others inverted over a flat stone or on the natural surface of the earth. This position, with the mouth downwards, is, perhaps, the most usual of the two. In some instances the bones were clearly enclosed, or wrapped, in a cloth before being placed in the urn. The place where the burning of the body has taken place is generally tolerably close to the spot on which the urn rests, or on which the heap of burnt bones has been piled up. Wherever the burning has taken place there is evidence of an immense amount of heat being used; the soil, for some distance below the surface, being in many places burned to a redness almost like brick. Remains of charcoal, the refuse of the funeral pyre, are very abundant, and in some instances I have found the lead ore, which occurs in veins in the limestone formation of Derbyshire, so completely smelted with the heat that it has run into the crevices among the soil and loose stones, and looks, when dug out, precisely like straggling roots of trees.

Is it too much to suppose that the discovery of lead may

be traced to the funeral pyre of our early forefathers? I think it not improbable that, the fact of seeing the liquid metal run from the fire as the ore which lay about became accidentally smelted, would give the people their first insight into the art of making lead—an art which we know was practised at a very early period in Derbyshire and other districts of this kingdom.* Pigs of lead of the Romano-British period, inscribed with the names of emperors and of legions, have occasionally been found; but much earlier than these are some cakes (if the term can be allowed) of lead which have evidently been cast in the saucer-shaped hollows of stones. Of these, which are purely British, some examples have fortunately been preserved.

But to resume. The positions I have spoken of in which the cinerary urns and heaps of burnt bones have been usually found, will be best understood by the accompanying engravings. The first (fig. 19) represents a section of a bar-

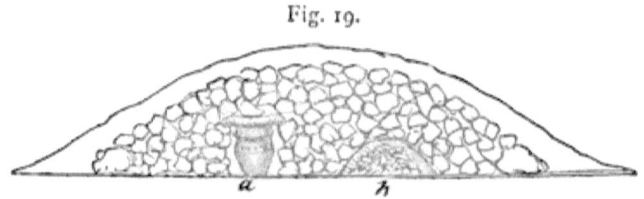

Fig. 19.

row in which, at *a*, is shown a sepulchral urn in an upright position, capped with a flat stone; and at *b* a heap of burnt bones piled up in the usual fashion, and first covered with earth and then with the loose stones of which the whole barrow was composed.

The next engraving (fig 20) again shows, within a cist,

---

* There are in Derbyshire lead mines worked at the present day which were worked, at all events, in the Romano-British period. Roman coins, fibulæ, and other remains are occasionally found in them.

INTERMENTS BY CREMATION. 33

in a barrow on Baslow Moor called "Hob Hurst's House,"*
two heaps of bones, the one simply collected together in a
small heap, and the other guarded by a row of small sand-

Fig. 20.

stone "boulders" all of which had been subjected to fire.
The next illustration (fig. 21) gives a section of the Flax
Dale barrow at Middleton by Youlgreave, which shows the

Fig. 21.

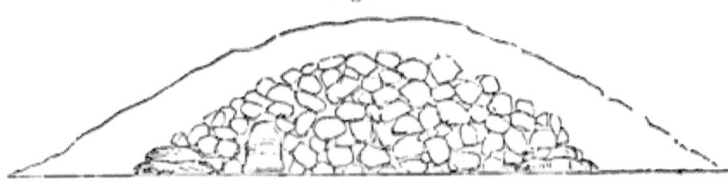

inverted position of the sepulchral urn. This barrow was
formed on a plan commonly adopted by the ancient
Britons, and will therefore serve as an example of mode
of construction as well as of the inverted position of the

* "Ten Years' Diggings."

urn. A circle of large rough stones was laid on the surface of the ground, marking the extent of the proposed mound. Within this the interments, whether in an urn or not, were placed, and the mound was then raised of stones to the required height, and afterwards covered to some thickness with earth, and thus the outer circle of the barrow was considerably extended, as will be seen by the engraving.

Another excellent example of the inverted position of the

Fig. 22.

sepulchral urns is here given (fig. 22) from one of the cists in Rolly-Low, near Wardlow. I have chosen it because, when found by Mr. Bateman, it had received a considerable fracture on one side, and thus showed the burnt bones which it contained, through the aperture. The urn was about sixteen inches in height and twelve inches in diameter, and was ornamented in the usual manner with indentations produced by a twisted thong. It was inverted over a deposit of calcined human bones, among which was a large red deer's horn, also calcined. The urn was so fragile as to be broken to pieces on removal.

In some urns discovered in Cambridgeshire, at Muttilow Hill, the Hon. R. C. Neville found that the calcined bones had been collected and wrapped in cloth before being placed in the urns. The contents of one of the urns he describes as " burnt human bones enveloped in a cloth, which, on looking into the vessel, gave them the appearance of being viewed through a yellow gauze veil, but which upon being touched dissolved into fine powder."* The urns were all inverted.

A somewhat peculiar feature of urn burial was discovered at Broughton, in Lincolnshire, where the urn containing the burnt bones was placed upright on the surface of the ground, and another urn, made to fit the mouth, inverted into it to form a cover.

In instances where the ashes of the dead have been collected from the funeral pyre, and laid in a skin or cloth before interment, the bone or bronze pins with which the " bundle" was fastened still remain, although, of course, the cloth itself has long since perished.

In other instances small stones have been placed around, and upon, the heap of burnt bones before raising the mound over the remains.

It is frequently found in barrows, where the interment has been by cremation, that there will be one or more deposits in cinerary urns, while in different parts of the mound, sometimes close by the urn, there will be small heaps of burnt bones without any urn. The probable solution of this is, that the simple heaps of bones were those of people who had been sacrificed at the death of the head of the family, and burned around him.

* Although I am describing the position in which the urns have been placed, it must not for a moment be supposed that they are often found in a perfect state, or in the position in which they have originally been placed. On the contrary, the urns are usually very much crushed, and not unfrequently, from pressure of the superincumbent mass of stones and earth, are found on their sides, and crushed flat.

It is a very frequent occurrence in barrows for the interments to be made in stone cists, and these, of course, vary both in size and in form, according to the nature of the spot chosen, and to the requirements of each particular case. The cists are usually formed of rough slabs of limestone, grit-stone, granite, or other material which the district offers, set up edgeways on the surface of the ground, so as to form a sort of irregular-square, rhomboidal, or other shaped compartment. In this the interment, whether of the body itself or of the urn containing the calcined bones, or of the calcined bones without an urn, has been made, and then the cist has been covered with one or more flat stones, over which the cairn of stones, or earth, or both, has been raised. Some barrows contain several such cists, in each of which a single, or in some instances a double, interment has been made. Excellent examples of these are afforded by the accompanying engravings, and by figs. 9, 10, 11, 12, 15, 17, 20, 28, and 29. Occasionally, when the natural surface of the ground was not sufficiently even or solid for the interment to be as conveniently made as might be wished, a flooring of rough slabs of stone was laid for the body to rest upon. This was the case in a barrow called "Long-Low," near Wetton, in the moorlands of Staffordshire (shown on fig. 24), which was opened by Mr. Carrington.

As there are some singular features connected with this

Fig. 23.

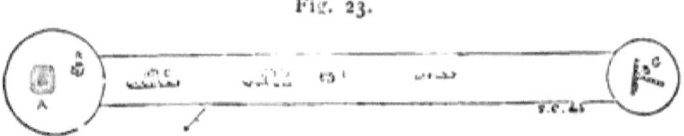

barrow, a detailed account of the mode of its construction becomes necessary. This very peculiar barrow had been thought, time out of mind, to be a "mine rake," and attempts have from time to time been made by lead miners to find a shaft, by removing certain portions of the mound—a shaft had, in fact, been sunk very nearly in the centre of the barrow. This rendered the operations of opening both difficult and laborious. Long-Low is what is usually denominated a "twin barrow," consisting of two circular mounds, connected by a bank, which altogether are 220 yards long. A plan of this barrow, drawn to scale, is shown on the accompanying engraving (fig. 23). The circular mound at the north-east end is thirty yards across and seven feet high in the centre, that at the south-west end not so large; the connecting bank at its base is fifteen yards wide, and where entire about six feet in height, with regular sloping sides where not mutilated. "The barrow runs in a straight line along the highest part of the land, a strong wall, separating the fields, is built over it lengthways, the stone for which, like other field walls in the vicinity, appears to have been procured from the bank of the tumulus, which, with the exception of some parts of the surface, is formed of large flat stones, which have evidently been procured in the immediate neighbourhood, where the surface of the land is lowered to a considerable extent. This is the only instance (as far as my experience goes) of a barrow being formed of stones got by quarrying, they being generally composed of such stones as are found on the surface of uncultivated land, which, owing to exposure to the atmosphere, have their angular points rounded. The strata in the neighbourhood of this barrow are but slightly consolidated, and are separated from each other by a thin seam of earthy matter, and abound with vertical cracks, so that it would not be a difficult task to dig out stones with sharpened stakes—the principal instruments, I presume, for such purposes in primitive times."

The internal construction of this cairn is singular. By making holes in various places along the bank, was found a low wall in the centre, built with large stones, which appears to be carried the whole length of the bank. Against this, large flat stones, with their tops reclining against the wall, are placed, thus leaving many vacancies, and showing an economical way of raising the mound at less expenditure both of labour and materials. The portions of this which have been laid bare are, with remains of interments, shown on the plan (fig. 23) at C, D, E, and F.

A large cist, or chamber, was discovered near the centre of the large mound. It was formed by four immense stones, inclosing an area six feet long, five feet wide, and about four feet deep. In all probability the capstone had been removed, as none was found. On the cist being cleared, was discovered a regular paved floor of limestone, entirely covered over with human bones, presenting a confused mass of the relics of humanity. The skeletons lying in the primitive position, crossed each other in all directions. They proved to be the remains of thirteen individuals, both males and females, varying from infancy to old age. The interior of this cist is shown on the accompanying engraving, and its position in the mound will be seen at A on the plan.

Fig. 24.

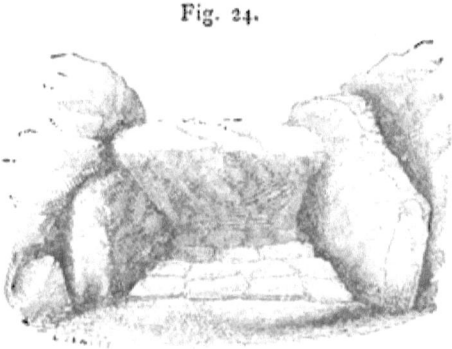

On the floor were found three arrow-heads of flint, wrought into beautiful thin leaf-shaped instruments, and many other calcined flakes of the same material; also bones of the ox, hog, deer, and dog. Not far from the cist, and near the surface, was found a skeleton minus the head, imbedded in gravel, rats' bones, and charcoal. On the floor some animal bones were found that had been burned, also neatly wrought arrow-heads and pieces of flint, and fragments of two human skulls. The point of a bone spear and a bone pin were found during our labours in this mound. "Another skeleton was found in the bank, crushed into small fragments; and where another grave had been made in the bank, for a secondary interment, the sides and bottom were found to have been burned to lime, which now resembled old mortar, to an extent that could not have been effected by an ordinary fire. It is not unusual to find small stones burnt to lime on the floors of barrows; in the present instance it had acquired a hardness almost equal to the stones, effected during a very long period by imbibing carbonic acid gas from the atmosphere, to which it had free access; pieces several inches thick were broken up intermixed with charcoal."

On one side of the cist two skulls lay close together, and

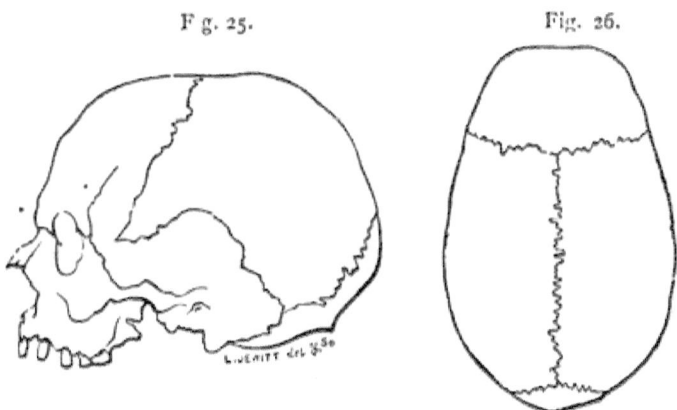

Fig. 25.   Fig. 26.

mixed up with a skeleton, the bones of which, in some

instances, crossed each other; in the centre lay the fine skull shown in figs. 25 and 26.* Two other remarkable skulls, one of a woman of about fifty years of age, and the other of a girl not more than seven years old, were also found.

When the mound at the other extremity of the "bank" was opened, calcined bones and animal remains only were found, but the singular construction of this portion of the barrow made ample compensation for the paucity of relics. It appeared that the longitudinal wall, noticed before, terminated

in the centre of this mound; and at its termination another and well-built wall was carried crossways at right angles with it (fig. 27), which was laid bare to the length of more than

* This skull has been most skilfully figured in "Crania Britannica," where it is carefully described and compared with other examples by Dr. Davis, who gives an admirable account of the discoveries at Long-Low, and of the characteristics of the different crania found there. Of the skull here shown Dr. Davis says it is "remarkably regular, narrow, and long; of good shape, medium thickness, and presenting few of the harsh peculiarities of the ancient British race; on the contrary, there is about it an air of slenderness and refinement. In some features it assimilates to the modern English cranium, although decidedly narrow, whilst its genuine and remote antiquity is determined by unquestioned evidence. It belongs, in an eminent degree, to the class of dolichocephalic skulls, and is the cranium of a man of about forty years of age."

half the diameter of the mound—it was three feet in height—the whole extent was not proved. From the centre of this wall, and forming a straight line with the longitudinal one, there was a row of thin moderately large stones, set on edge, by the ends being set in the soil that formed the floor of the mound; these were placed with their edges close together, and occasionally in two or three ranks, as if for better support in an upright position. They were from 1½ to 2 feet in height, and were extended from the wall to the length of five yards. The burnt bones were found in the west angle formed by the cross wall and the upright stones, as shown in the engraving. It appeared as though the bones had originally been deposited near the surface, as they were now found in the interstices betwixt the stones, from near the top to the bottom. This mound was formed of large stones, like the other parts, reared against each other all around, with their tops inclining towards the centre.

A tolerably good cist, formed of rough masses of stone surrounding the body, is the one engraved on fig. 28, from

Middleton. This cist contained the skeleton of a woman, lying on her left side, in a partially contracted position. Above her lay the remains of an infant, and about her neck were the beads of a remarkably fine necklace of jet.

Another good example of a stone cist is the next (fig.

29), from Liff's-Low. The cist was formed of eight large slabs of rough limestone, set edgewise; and formed a chamber of very compact and almost octagonal form. It contained the skeleton of a man, lying in a partly contracted posture on his left side, the face looking to the west. Behind the knees was placed a hammer-head, formed of the lower part of the horn of the red deer. One end is rounded

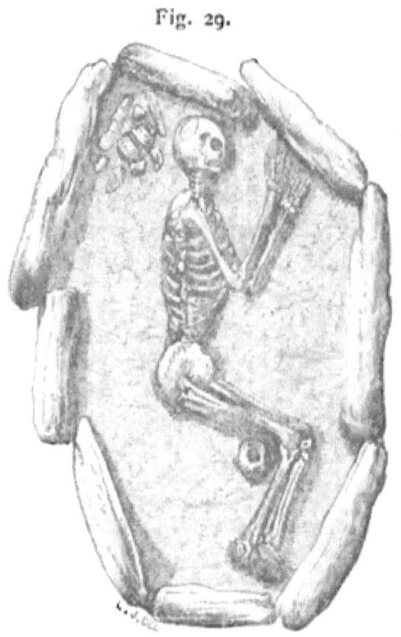

Fig. 29.

and polished, and the other knotched across, somewhat, (to use a homely comparison) like what is usually called a "wafer-seal." Behind the head were a number of miscellaneous but highly interesting articles, showing, as they did after the lapse of so many centuries, that " the savage Briton, reposing in this cairn, had cultivated the art of making war amongst the inhabitants of the forest in preference to molesting his fellow savages; as almost the first observed articles were a pair of enormous tusks of the wild boar, trophies of some, perhaps his last, sylvan triumph. Next came two arrow-heads of flint, delicately chipped, and of unusual form; two spear-heads of the same material; two flint knives, polished on the edge, one of them serrated on the back in order to serve as a saw; and numerous pieces of flint of indescribable form and use, which, together with all the flint instruments enumerated above, seem to have undergone a partial calcination, being gray, tinted with various shades of blue and pink. With these articles were found three pieces of red ochre, the rouge of these unsophisticated huntsmen, which, even now, on being wetted, impart a bright red colour to the skin, which is by no means easy to discharge." With these articles lay a small urn of unique form.

On fig. 30 is shown a remarkably pretty cist, formed of

Fig. 30.

four upright stones, supporting a capstone. It contained a vase of good form.

Pit interments are occasionally met with, but are very rare. One instance will suffice: it is that at Craike Hill,

near Fimber, Yorkshire, and was opened by Mr. Mortimer.*
In it no less than four interments were made, one above
another, in a pit or grave covered over by a mound. The
two lower and the upper were skeletons in the usual contracted positions; the other a heap of calcined bones, with
a fine food vessel. Near the upper skeleton was another
heap of burnt bones and another food vessel. With one of
the skeletons, that of a female, was found a splendid necklace of jet, and a drinking cup of elaborate design. These
are all engraved in the present volume.

Burials in tree-coffins, of various periods, have been
occasionally met with in grave-mounds, and a few words
concerning them may here be introduced. One of the most
interesting was found recently by that indefatigable antiquary, the Rev. Canon Greenwell, at Scale House, in
Yorkshire.† The barrow was about thirty feet in diameter,
and five feet in height, and was surrounded by a circle of
soil at the base. It was entirely composed of soil, interspersed here and there with fragments of charcoal, firmly
compacted. On the top, for a space of about six or seven
feet in diameter, a covering of flattish stones was laid just
below the surface. On digging down at this spot it was
found that a hollow had been made in the natural surface,
that had been filled up with soil, upon which had been
placed a few stones and then a coffin, constructed of the
trunk of a small oak tree. This primitive coffin was laid
north and south, the thicker end, which no doubt contained
the head of the corpse, towards the south; which was also
the case in the Gristhorpe barrow. The oaken trunk, or
tree-coffin, was seven feet three inches in length, and one
foot eleven inches in diameter, at the spot in which it was
measured. The Gristhorpe coffin "is seven feet and a

---

* Described in the " Reliquary," vol. ix.
† For a full account of this discovery see the " Reliquary," vol. vi. page 1.

half long, and three feet three inches broad." Above the coffin the soil was finer, and upon this finer stratum was situated a layer of dark matter a good deal burned, and containing pieces of charcoal. Over the whole was a covering of the ordinary compacted soil of which the barrow was composed. The body inhumed in this tree-coffin, and not burned, had gone totally to decay, leaving only a whitish unctuous matter behind. This substance was no doubt *adipocire*, the production of which is to be accounted for by the extreme wetness of the barrow. Before interment the corpse had been clothed, or wrapped, *from head to foot* in a woollen fabric,* a specimen of which is represented in the following figure :—

Fig. 31.

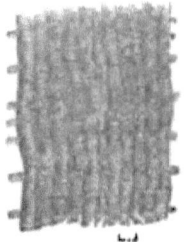

There were no flint chippings discovered in the soil of which the barrow was composed, or other object, and nothing else was contained in the tree-coffin, save the body in its wrappings. Some pieces of a bright black substance

* This woollen cloth must be regarded as a *woven* texture, but whether it were woven in so artificial a machine as a loom may be questioned. A great variety of contrivances have been used for weaving, *i.e.*, crossing alternately threads passed in opposite directions, the warp and the woof, by what are called *savage* races. Still it is not at all improbable that a people so advanced in pastoral habits, possessed some machine for weaving, bearing a relation to a primitive loom. Both warp and woof are composed, as might be expected, of a simply spun thread of one strand. Perforated stones are found in British and Danish barrows, and perforated pieces of earthenware in the Swiss Lake villages, even of the stone period, which are regarded as spindle-whorls.

like pitch, which appeared to have been placed on the inside of the coffin, on examination are found to be composed of carbon and oxide of iron.

It is most unfortunate that this curious and interesting barrow had been previously opened at the top. By this proceeding the tree-coffin had been broken through, and its contents disturbed about the middle. And it is also much to be regretted that the barrow was saturated with moisture, which had percolated into the coffin, carrying the soil with it. By this means all the contents of the barrow, save the *adipocire* of the body itself, including both the tree-coffin and the woollen garment, had acquired a rottenness which precluded the recovery of anything more than mere fragments. Those of the woollen dress were so filled with particles of soil, and at the same time so tender, as to admit of being reclaimed only in a very imperfect manner. The woollen cloth, which went from head to foot, there is no doubt had been loosely wrapped round the body, in the manner of a shroud, not swathed like an Egyptian mummy, so that " the fabric filled the whole of the inside of the coffin from end to end." Hence, as is confirmed by the barrows opened in Jutland, there is every reason to infer that it was the ordinary woollen dress of the individual interred in the tumulus, who must have held such a position in society as to ensure these great attentions to his remains.

" In many ancient British barrows marks of the garments of the deceased have been discovered, in which the body appears to have been wrapped before interment. Indications of skin dresses are seen early, and after these, in the bronze and iron periods, where the rust of weapons has retained impressions of such grave-clothes, tissues of linen and woollen appear. Mr. Bateman met with signs of such textures, and in the case of the Tosson cists, in Northumberland, from one of which the skull of plate 54 of the

Crania Britannica was derived, an iron spear-head was found in one of them, and there were signs of two fabrics of cloth impressed upon the oxidized surface of this relic.

"Again, British barrows have been opened containing tree-coffins, in which the remains have been inhearsed. The celebrated Gristhorpe barrow, the skeleton from which is preserved in the Scarborough Museum, and of the skull of which there is a fine engraving, plate 52, and a careful description by Dr. Thurnam, in the work just named, offers an instance of a tree-coffin formed of a split oak of small girth. In this case, the body had been wrapped in the skin of some animal having soft hair. The interment had belonged to the ancient British late stone or the bronze period. The coffin contained three flakes, or rude implements of flint, as well as objects made of bronze and bone. In the course of the description alluded to, there are references to many other examples of coffins hollowed out of solid trunks, oaken and tree-coffins. These appear to belong to very different periods, extending from the ancient British to early Saxon, and perhaps Christian times. That called the "King Barrow," at Stowborough, in Dorsetshire, contained an oaken tree-coffin with the body in an envelope of deer-skins. It is said, that more recently a barrow opened in the wolds of Yorkshire offered fragments of an oaken coffin, together with the remains of a British urn. Also, at Wath, in the North Riding of Yorkshire, in an oaken coffin an urn was found of the later British type, the whole being enclosed in a barrow.

The Gristhorpe coffin, shown on figs. 32 and 33, consisted of the trunk of a large oak, roughly hewn, and split into two portions. The markings seemed to indicate that it had been hollowed with chisels of flint; but that the tree had been cut down with a much larger tool, the marks being such as would be made by a stone hatchet. It is seven and a half feet long and three feet three inches broad. In the bottom is a hole three inches in length. The lid was kept in place by the uneven fracture of the wood. The back was in good preservation, with its coatings of lichens distinct. At the narrow end of the lid, cut in the bark, was a sort of leaf-shaped knob, perhaps intended for a handle. The objects found in the coffin alluded to above are shown on figs. 34 and 35. In these engravings, Nos. 1, 2, and 6 are flakes of flint. The first has been slightly chipped at the

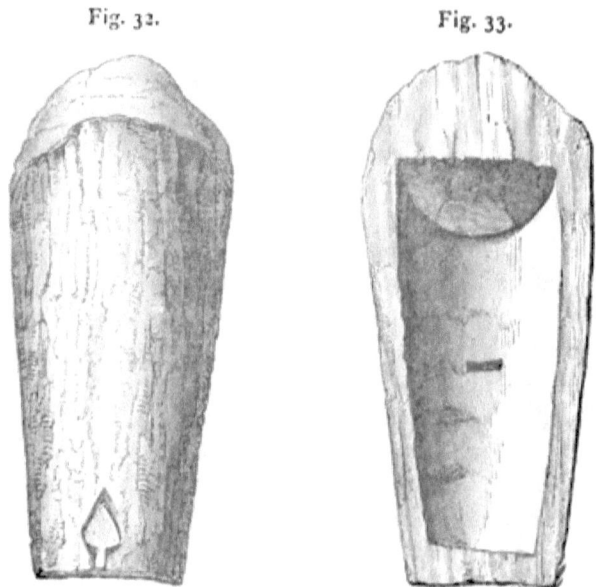

Fig. 32.   Fig. 33.

edge, but the others are simply split off from the native flint. No. 5 is a bronze dagger, three and a half inches long, but much corroded—the two rivets showing that the handle was not of much thickness. No. 4, no doubt the top of such handle, is a disc of bone, polished, and of oval shape with perforations on either side for the pins by which it was fastened. No. 8 is a small implement of wood, with a rounded head, and flattened on one side to about half its length. No. 3 is the fragment of a ring of horn—a fastening, perhaps, of the dress. On the lower part of the breast was an ornament of very brittle material, in the form of a rosette, with two loose ends. By the side was a shallow

Fig. 34.   Fig. 35.

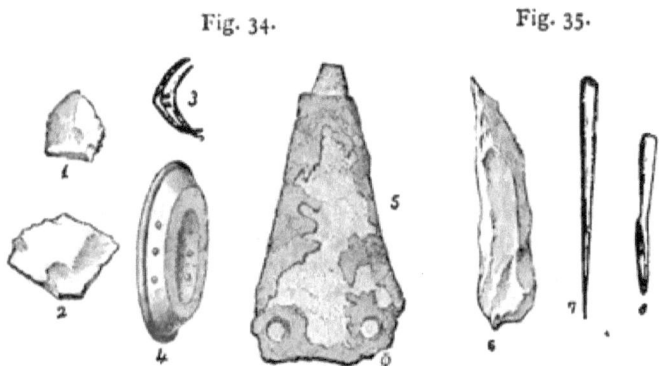

basket, about six inches in diameter, formed of bark, curiously stitched with the sinews of animals; at the bottom were decomposed remains, perhaps of food. There was also a quantity of vegetable substance, mixed with lanceolate foilage, supposed to be that of the mistletoe.

Fig. 36.

Another form—which may be called the "boat shape" —of tree-coffin is here shown for the purpose of comparison.

CHAPTER IV.

Ancient British or Celtic Period—Sepulchral Chambers of stone—Cromlechs—Chambered Tumuli—New Grange and Dowth—The Channel Islands—Wieland Smith's Cave, and others—Stone Circles—For what purpose formed—Formation of Grave-mounds—Varieties of Stone Circles—Examples of different kinds—Arbor Low,-etc.

ONE of the most important classes of barrows is that which contains sepulchral chambers of stone ; not the simple cists which have been spoken of in the preceding chapter, but of a larger, more complicated, or colossal character. Mounds of this description exist, to more or less extent, in different districts. In most instances the mound itself, *i.e.*, the earth or loose stones of which the superincumbent mound was composed, has been removed, and the gigantic sepulchral chamber alone left standing. In many instances the mounds have been removed for the sake of the soil of which they were formed, or for the purpose of levelling the ground in the destructive march of agricultural progress. In many cases, however, they have doubtless been removed in the hope of finding treasure beneath; it being a common belief that immense stores of gold—in one instance the popular belief was that a "coach of gold" was buried beneath—were there for digging for. Where the mounds have been removed, and the colossal megalithic structures allowed to remain, they have an imposing and solemn appearance, and seem almost to excuse the play of imagination indulged in by our early antiquaries in naming them Cromlechs, and in giving to them a false interest by making them out to be "Druids' altars" —altars on which the Druids made their sacrifices. These same authorities have, indeed, gone so far in their inventions as to affirm, that when the capstone was lower on one side than another, as must necessarily frequently be the case, it was so constructed that the blood of the victims might run off in that direction, and be caught by the priests; that some of the naturally formed hollows in the stones were scooped out by hand to receive the heart and hold its blood for the highest purposes; and that when the cromlech was a double one, the larger was used for the sacrifice, and the smaller for the Arch-Druid himself whilst sacrificing.

Researches which have been made in recent times show the absurdity of all this, and prove beyond doubt that the cromlechs are neither more nor less than sepulchral chambers denuded of their mounds. In several instances they have been found intact, and, these mounds being excavated, have been brought to light in a perfect state. These instances have occurred in Cornwall, in Derbyshire, and in other districts of England, as well as in the Channel Islands and elsewhere. One instance is that of the Lanyon cromlech in Cornwall. It seems that some seventy years ago "the farmer" to whom the ground belonged cast a longing eye on what appeared to be an immense heap of rich mould, and he resolved to cart it away and spread it over his fields. Accordingly he commenced operations, his men day after day digging away at the mound, and carting the soil off to the fields. By the time some hundred cart loads or so had been removed the men came to a large stone, which defied their efforts at removal, and, not knowing what it might be, or what it might lead to, they went on removing the surrounding earth, and gradually cleared, on all its sides, the majestic cromlech which is now one of the prides of Cornwall. This highly interesting chamber contained a heap of broken urns and human bones. This "Lanyon cromlech," a view of which is given on fig. 37, consists now of three immense upright stones, on which rests an enormous capstone, measuring about eighteen and

Fig. 37.

a half feet in length and about nine feet in width, and is computed to weigh above fifteen tons. How such stones were raised and placed on the rough upright stone supports

Fig. 38.

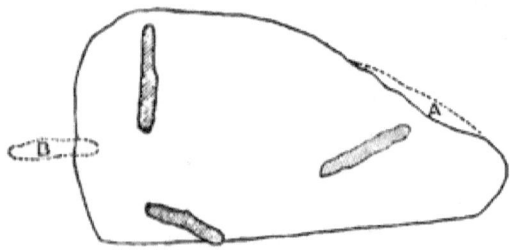

which had been prepared for them is almost beyond comprehension, when it is recollected that they were raised by a people who were devoid of machinery.

> "The heart,
> Aching with thoughts of human littleness,
> Asks, without hope of knowing, whose the strength
> That poised thee here."

This cromlech when first uncovered consisted of *four* upright stones, on which rested the capstone. In 1815, during a tremendous storm, the capstone and one of the supports were thrown down. In 1824 the capstone was replaced, under the superintendence of Lieut. Goldsmith, R.N., and at this time a piece was broken off at A. The fourth upright stone was not replaced, having been broken when thrown down. Fig. 37 shows the cromlech as replaced. Fig. 38 is a plan of it, showing the uprights and the capstone. The large outline is the capstone, the part marked A being the part broken off; the shaded parts are the present three uprights; and B the fourth upright, broken.

Fig. 39.

Kits Cotty House, in Kent; the Chun cromlech, in Corn-

Fig. 40.

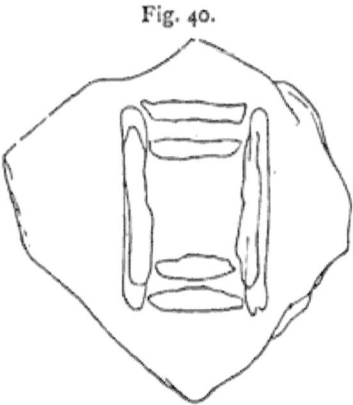

wall (figs. 39 and 40)—the covering stone of which is calculated to weigh twenty tons; the Molfra cromlech, in the same county, which consists of a compact cist closed on three sides and open on the fourth; the Zenor cromlech; the Plas Newydd cromlech, and many others which it is not necessary to enumerate, are all of the same class. The Plas Newydd (fig. 41) is a double cromlech, the two chambers being close together, end to end. The capstone of the largest, which is about twelve feet in length by ten

feet in breadth, originally rested on seven stones, two of which have disappeared. The two erections undoubtedly were originally covered with a single mound.

At Minninglow, in Derbyshire, erections of this kind occur, but, not being denuded of their mounds, are still partially buried. The mound is of large size. Under the centre and in four places in the area of the circle are large cists, which if cleared from the earth would be fine cromlechs of precisely the same form as those just described. They are formed of large slabs of the limestone of the district, placed upright on the ground, and are covered with immense capstones of the same material. All these chambers had contained interments. The accompanying plan (fig. 42) of some of these cists gives the situation of the stones forming the sides of the large chamber; of the passage leading to it; of the slabs which closed its entrance; and of the covers or capstones. The chamber is rather more than five feet in height, and the largest cap-

Fig. 42.

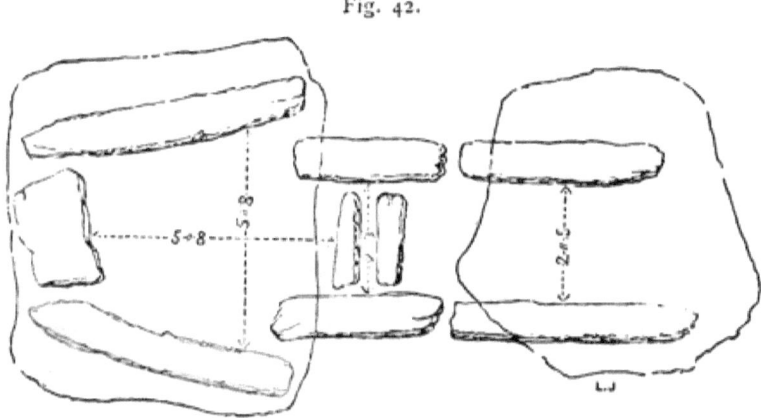

stone about seven feet square, and of great thickness. A kind of wall similar to those which have been found to encircle some of the Etruscan tumuli, forms the circle of this mound, which rises to a height of more than fifteen feet from the surface of the ground.*

The general arrangement of this example will be seen to bear an analogy to the Plas Newydd and others spoken of, and shows by what an easy transition the building of galleries, or a series of chambers for family tombs, in these large mounds, would be arrived at. Of this kind some very large examples exist in Ireland, and in the Channel Islands, as well as in various parts of England.

* It is worthy of remark, that this noble mound, with its very early interments, has been made a place of sepulture in more recent times, many Roman coins and remains of that period having been found there.

One of the most important in size, as well as in general interest, is the one at New Grange, county Meath. "The cairn, which even in its present ruinous condition measures about seventy feet in height, and is nearly three hundred feet in diameter, from a little distance presents the appearance of a grassy hill, partly wooded; but upon examination the coating of earth is found to be altogether superficial, and in several places the stones, of which the hill is entirely

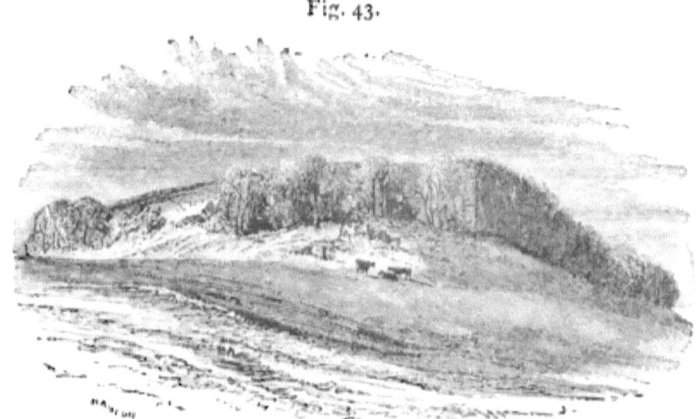

Fig. 43.

composed, are laid bare. A circle of enormous stones, of which eleven remain above ground,* originally encircled its base. The opening (of which an engraving is shown on fig. 44) was accidentally discovered about the year 1699, by labouring men employed in the removal of stones for the repair of a road. The gallery, of which it is the external entrance, extends in a direction nearly north and south, and communicates with a chamber, or cave, nearly in the

* These immense monoliths have originally, it is estimated, been upwards of thirty in number, and to have been placed probably ten yards apart. The largest remaining stone stands between eight and nine feet above the ground, and is seventeen feet in circumference. It is estimated to weigh upwards of seven tons. Several of the stones have entirely disappeared, of others fragments remain scattered about.

centre of the mound. This gallery, which measures in length about fifty feet, is at its entrance from the exterior about four feet in height, in breadth at the top three feet two inches, and at the base three feet five inches. These dimensions it retains, except in one or two places, where the stones appear to have been forced from their original

Fig. 44.

position, for a distance of twenty-one feet from the external entrance. Thence towards the interior its sides gradually increase, and its height where it forms the chamber is eighteen feet. Enormous blocks of stone, apparently water-worn, and supposed to have been brought from the mouth of the Boyne, form the sides of the passage; and it is roofed with similar stones. The ground plan of the chamber is cruciform; the head and arms of the cross being formed by three recesses, one placed directly fronting the entrance, the others east and west, and each containing a basin of

Fig. 45.

granite. The sides of these recesses are composed of

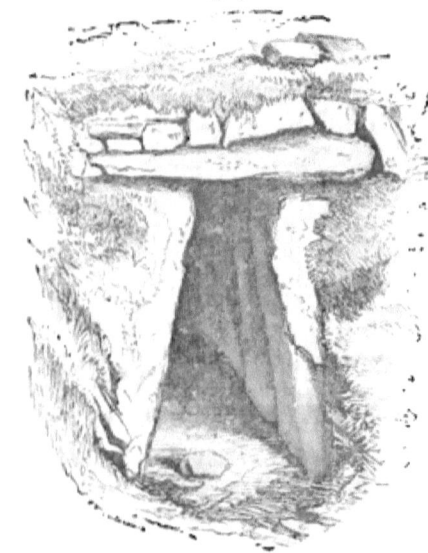
Fig. 46.

immense blocks of stone, several of which bear a great

CROMLECHS.   59

variety of carvings."* In front of the entrance (fig. 44) will be seen one of these carved stones.

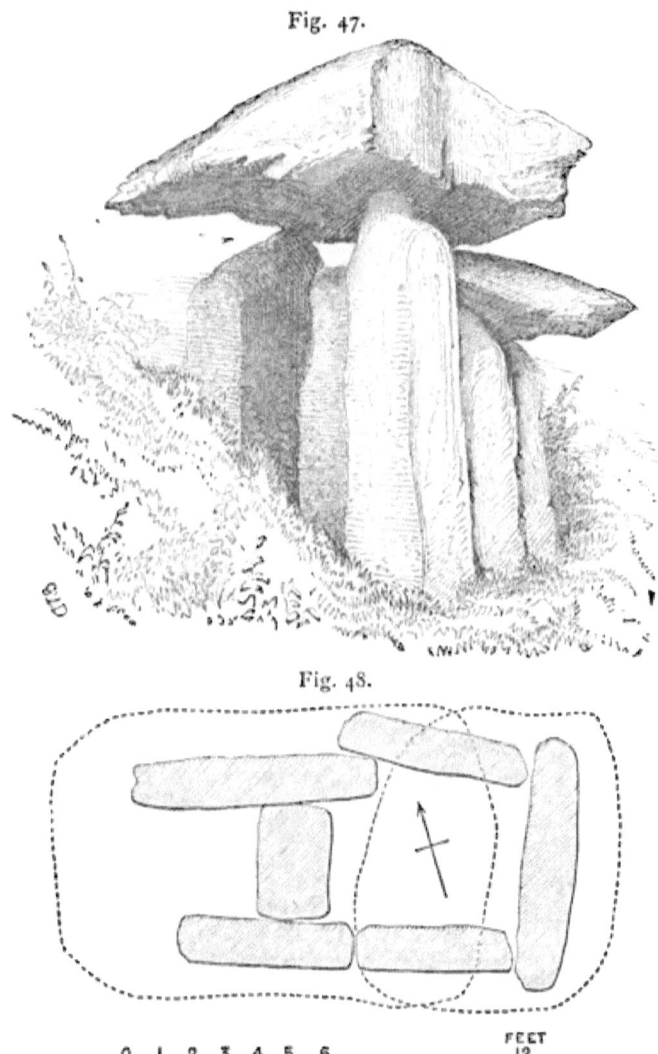

Fig. 47.

Fig. 48.

At Dowth and Nowth (Dubhath and Cnobh), very similar

* For an excellent notice of this and other remains, the reader is re-

chambered tumuli exist, the former of which is also re-

Fig. 49.

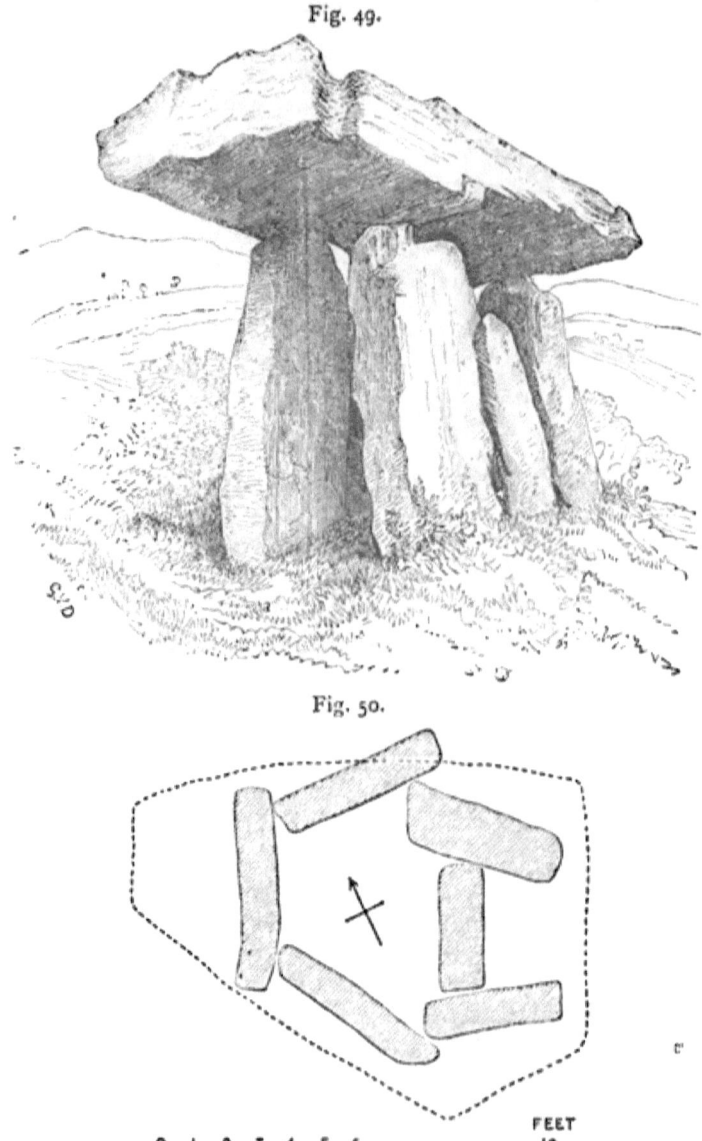

Fig. 50.

markable for its sculptural stones, which bear a strong resemblance to those at New Grange. The Cairn of Dowth here engraved (fig. 45), is of immense size, and contains a cruciform chamber similar to that at New Grange, with a passage twenty-seven feet in length, composed—as was the chamber—of enormous stones. On some of the stones were carvings and Oghams. The mouth of the passage leading to the cruciform chamber is shown on fig. 46.

Other excellent examples of Irish cromlechs and chambers are those at Monasterboise ("Calliagh Dirras House"); Drumloghan (full of Oghams); Kells; Knockeen (figs. 47 and 48); where the right supporting stones are six in number,

Fig. 51.

and arranged rectangularly, so as to form a distinct chamber

ferred to Mr. W. F. Wakeman's "Handbook of Irish Antiquities,"—the best and most compact little work on the subject which has been issued, and one which will be found extremely useful to the archæological student—to which I am indebted for some of the accompanying engravings.

at the S.E. end, the large covering stone being 12 feet inches by 8 feet, and weighing about four tons, and the

Fig. 52.

smaller one about half that size; Gaulstown (figs. 49 and 50, the inner chamber of which measures 7 feet by 6 feet

Fig. 53.

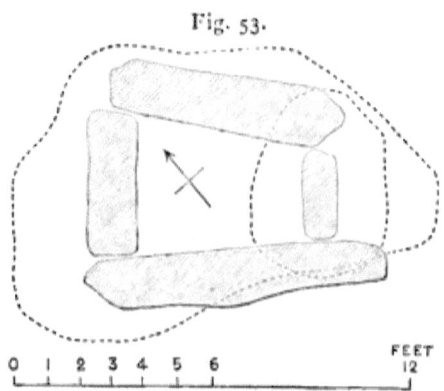

4 inches, and is seven feet in height); Ballynageerah (figs. 51, 52, and 53), the capstone of which is cleverly and

curiously poised on two only of the upright stones, as will be seen by the engravings;* Howth, Shandanagh, Brennanstown, Glencullen, Kilternan, Mount Brown, Rathkenny, Mount Venus, and Knock Mary, Phœnix Park, as well as at many other places.

In the Channel Islands the indefatigable and laudable researches of Mr. Lukis show that the galleried stone chambers of the tumuli in that district had been used by successive

Fig. 54.

generations for many ages. One of the most important of these is the gigantic chambered burial place, surrounded by a stone circle, at L'Ancresse, in Guernsey. In this, " five large capstones are seen rising above the sandy embankment which surrounds the place; these rest on the props beneath, and the whole catacomb is surrounded by a circle of upright stones of different dimensions. The length of

---

* For the loan of these seven engravings I am indebted to the Council of the "Historical and Archæological Association of Ireland," (formerly the "Kilkenny and South-east of Ireland Archæological Society,") in whose journal—one of the most valuable of antiquarian publications—they have appeared. This Association is one of the most useful that has ever been established, and deserves the best support, not only of Irish, but of English antiquaries.

the cromlech is 41 feet from west to east, and about 17 feet from north to south, on the exterior of the stones. At the eastern entrance the remains of a smaller chamber is still seen; it consisted of three or four capstones, and was about seven feet in length, but evidently within the outer circle of stones.* In a careful examination made by Mr. Lukis, many highly interesting features were brought to light, of which he has given an excellent account in the "Archæological Journal,"† to which the reader cannot do better than refer. The engravings there given, show the interiors of some of the chambers, with their deposits *in situ*, and exhibit some of the highly interesting relics found during the excavations. The pottery was of a totally distinct character from that of the Celtic period found in England, some of the forms being of what are usually considered the Anglo-Saxon type, and are the result of the use of these chambers by successive generations, as already named.

Another of the more remarkable structures of the Channel

Fig. 55.

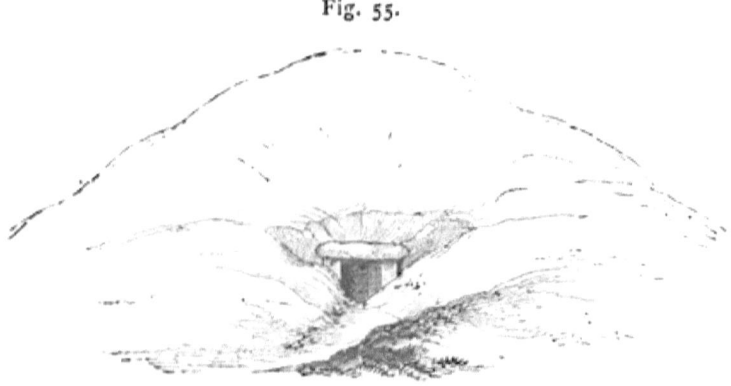

Islands is that of Gavr Innis, in the Morbihan, Brittany. The tumulus is about thirty feet high, and its circumference

* F. C. Lukis. † Vol. i. p. 142.

at the base about 300 feet. The cromlech is entered from

Fig. 56.

the south end (fig. 55), fig. 56 being the opening on the north,

Fig. 57.

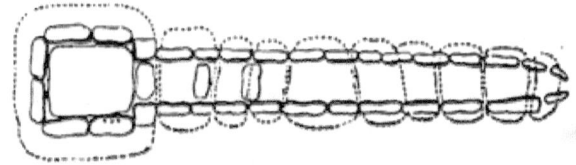

and consists of 14 upright stones on the east side, 13 on

Fig. 58.

the west, and 2 on the north, supporting, in all, 10 capstones. In general features it bears a strong resemblance to those at New Grange, Dowth, and other places. The

Fig. 59.

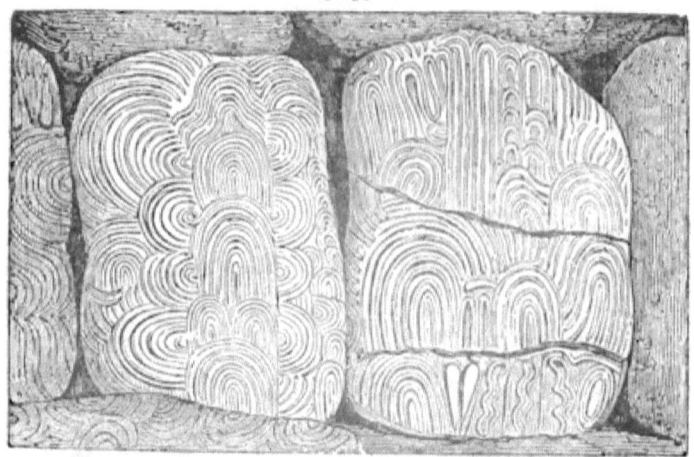

remarkable feature of this chambered tumulus is that the

Fig. 60.

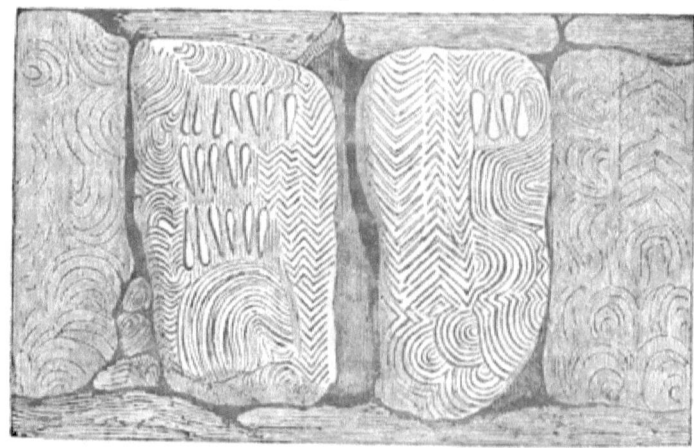

stones composing the passage are for the most part sculptured in lines and patterns, which have been described as very similar to the patterns tattooed on their faces and bodies by the New Zealanders. Examples of these will be seen on the accompanying engravings, which exhibit some of the more marked and distinct of the patterns noticed and copied by Mr. Lukis, in his examination of this mound, and

Fig. 61.

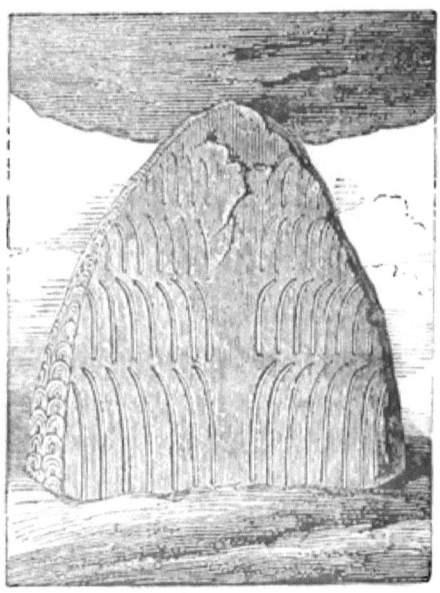

described by him in the journal of the British Archæological Association, to the Council of which I am indebted for these illustrations.

In England, besides those already named, and others, "Wayland," or "Wieland Smith's Cave," at Ashbury, in Berkshire; one at Stoney Littleton, near Wellow, in Somersetshire; the "Five Wells," at Taddington, in Derbyshire; and one or two others in the same county, as well as in other places, are the most important. The annexed woodcut, fig. 62, exhibits a section of the chambered tumulus at Stoney Littleton, and fig. 63 is a ground plan of the same.

"The entrance was on the north-west side, where a stone upwards of seven feet long and three and a half feet wide, supported by two others, left a square aperture of about four

Fig. 62.

feet high, which had been closed by another large stone. This entrance led to a long passage or avenue, extending in the direction from north-west to south-east forty-seven feet

Fig. 63.

six inches, and varying in breadth. There were three transepts, or recesses, on each side. The side walls were formed of thin laminæ of stone piled closely together without cement, and a rude kind of arched roof made of stones so placed as to overlap each other. When the large stones in the side walls did not join, the interstices were filled up with layers of small ones."* Interments had evidently

* T. Wright.

been made in each of these chambers, some by cremation, and others by inhumation, but the bones were scattered about, the result of previous rifling of its contents. One urn was found.

Fig. 64.

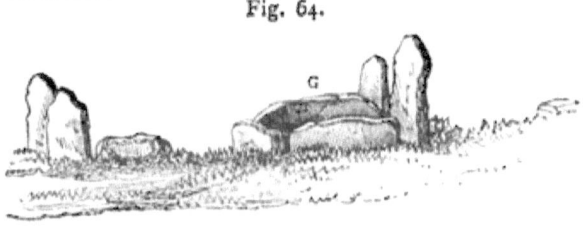

The chambered tumulus, called the "Five Wells," near

Fig. 65.

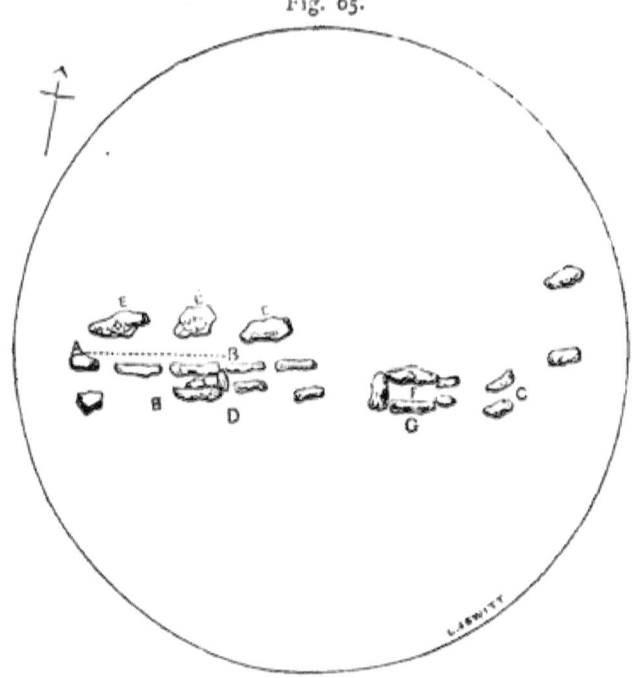

Taddington, of which an engraving is here given (fig. 64), has been a mound of large size, and the chambers and passages, or gallery, have been extensive. A plan of this tumulus is given in fig. 65. The "Five Wells" tumulus consists of two vaults or chambers, situated near the centre of a cairn (which is about thirty yards in diameter), each approached by a separate gallery or avenue, formed by large limestones standing edgeways, extending through the tumulus, respectively in a south-east and north-west direction. These chambers are marked B and G on the plan, G being the cist engraved on fig. 64. E E E are stones supposed to be the capstones thrown down. Another five-chambered tumulus in the same county is called Ringham-Low, which has many interesting remains.

Another extremely important mound of this description is the one at Uley, in Gloucestershire, of which an able account has been written by Dr. Thurnham.* The mound is about 120 feet in length, 85 feet in its greatest breadth, and about 10 feet in height. It is higher and broader at its east end than elsewhere. The entrance at the east end is a trilithon, formed by a large flat stone upwards of eight feet in length, and four and a half in depth, and supported by two upright stones which face each other, so as to leave a space of about two and a half feet between the lower edge of the large stone and the natural ground. Entering this, a gallery appears, running from east to west, about twenty-two feet in length, four and a half in average width, and five in height; the sides formed of large slabs of stone, set edgeways, the spaces between being filled in with smaller stones. The roof is formed, as usual, of flat slabs, laid across and resting on the side-slabs. There are two smaller chambers on one side, and there is evidence of two others having existed on the other side. Several skeletons were found in this fine tumulus when it was opened, many years ago.

It will have been noticed that circles of stones surround-

* "Archæological Journal," vol xi., p. 315.

ing grave-mounds have frequently been named in this and the preceding chapter. It will, therefore, be well to devote a few lines to these interesting remains.

Circles of stone of one kind or other are not unfrequently to be noticed in various parts of the kingdom, and they vary as much in their size and in their character as they do in their other features. The bases of grave-mounds were frequently defined by these circles, and sometimes by a shallow fosse, and occasionally by a combination of both. To this circumstance the origin of many of the circles of stones remaining to this day are to be traced; while others of a far larger construction, and of a totally different character, such as those of Stonehenge, Abury, Rollrich, and, probably, Arbor-Low, have been formed for totally different purposes. With these larger ones, except in so far as they are connected with sepulchral tumuli, I have in my present work but little to do. Of the smaller ones, those which have surrounded grave-mounds, I will now proceed to give some particulars.

Excavations into various grave-mounds have proved beyond doubt the fact that, in many instances, when an interment was made, the size of the proposed cairn to be raised over the remains was marked by a circle of stones laid on the surface of the ground, or inclining inwards, or set upright in the earth. The stones were then piled up within this enclosure, till the whole size and altitude of the mound was reached. In the case of the Flax Dale barrow, this mode of construction is shown in the next engraving (fig. 66). A circle of large flat stones was placed upon the surface of the earth, around the interment (which in this case consisted of calcined bones, in urns and without), and upon these a second course of stones was placed. The mound was then raised in the manner indicated in fig. 4, and over this a thick layer of earth was laid, which increased both the circumference and the altitude of the barrow. To render this crust more compact, fires were evidently lit on the circumference of the circle, which had

Fig. 66.

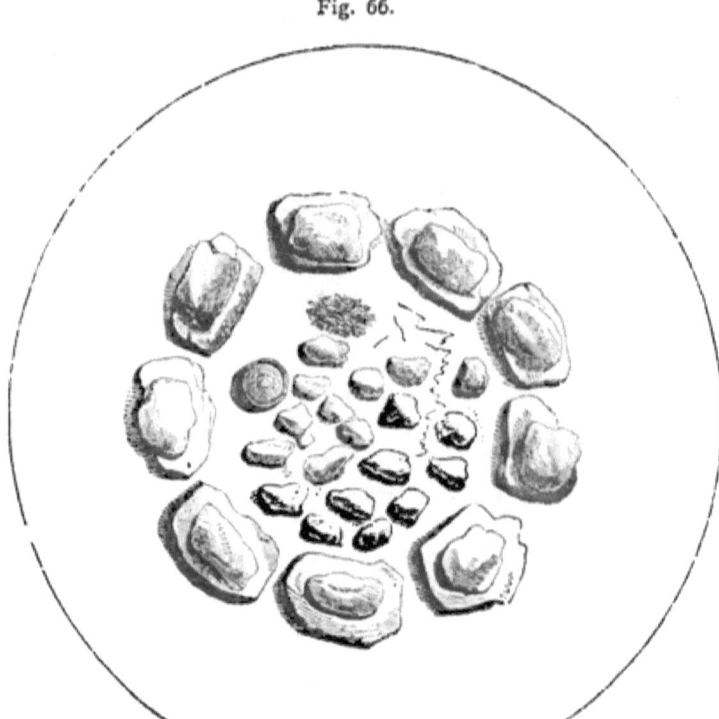

the effect, by burning the soil, of hardening it, and making it in some cases almost of the consistency of brick.

An example of the second mode of construction I have indicated is at Elk-Low, a section of which is given on fig. 67. The barrow has a depression running around its upper surface, something like an elevated fosse, as will be seen in the section. The interments were made on the natural surface of the ground, where, in the centre, lay a skeleton, on its right side, in a contracted position, with its head resting on a piece of limestone which was placed as a pillow. Other skeletons were also found, as was likewise an interment of burnt bones, and some flint and stone

Fig. 67.

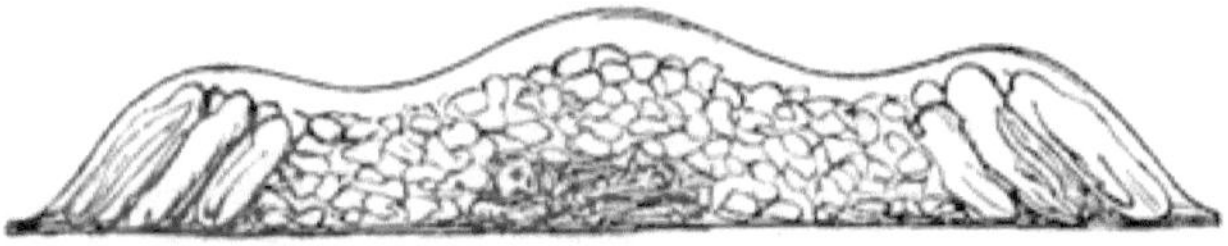

instruments. The outer circle was constructed of very large stones inclining inwards, and covered with small stones and earth, and thus forming an extremely durable mound.

Both of these examples, if denuded of their mounds, would form striking and very perfect stone circles, and would be among the best remaining examples of small "Druidical circles," as they are commonly called.

Where the circles have been formed of upright stones, they have not, certainly, always been covered with the mound, but have formed a kind of ring fence, a sort of sacred enclosure, around the barrow. A great number of examples of this kind exist in different districts, and will easily be recognised by the zealous archæologist. The circle next shown, on figs. 68 and 69, is that on Stanton Moor,

Fig. 68.

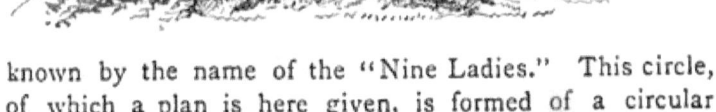

known by the name of the "Nine Ladies." This circle, of which a plan is here given, is formed of a circular mound of earth, on which the upright stones are placed.

It is about thirty-six feet in diameter. It has formerly consisted of a larger number of stones; those that are now

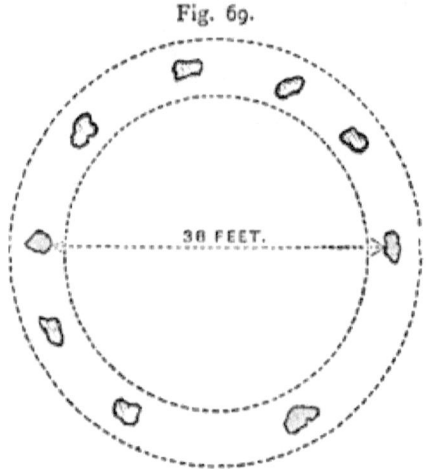

Fig. 69.

remaining being at irregular distances, varying from eight to nineteen feet apart. In the centre are the remains of a rifled sepulchral mound.

Another circle, bearing the same name, "The Nine Ladies," is on Hartle Moor, but of this only four stones are now remaining. It has undoubtedly been a sepulchral mound, encircled by upright stones. On other parts of these moors other circles have existed, or still exist, which have, by excavations, been proved to have enclosed sepulchral deposits.

On Brassington Moor, near a fine chambered tumulus, now unfortunately destroyed, existed two similar circles, the one thirty-nine, and the other twenty-two, feet in diameter. On Leam Moor, too, circles are known to have existed, surrounding interments. On Eyam Moor circles of this kind, encircling sepulchral mounds, exist. One of these is about a hundred feet in diameter, and is, like the "Nine Ladies" on Stanton Moor, formed of a circular mound of earth, on which the stones are placed. Only ten of the stones remain *in situ*. In the centre a cist was discovered many years ago. Other circles occur in the same county, on Abney Moor, on Froggat Edge, on the East Moor, on Hathersage Moor, and in other localities.

On Dartmoor, in Devonshire, many circles yet remain, as they do also in Cornwall and in other counties. Mr. Blight, who has paid a vast deal of attention to the antiquities of his native county, Cornwall, has collected together many data concerning these structures, which tend to throw much light upon their modes of construction, as well as uses. To his researches I am indebted for much of the following information regarding the Cornish circles, and also for the diagrams which illustrate it. Upright stones were, as in the case of the ring fences already named, placed at tolerably regular intervals around the barrow, either on the natural surface of the ground or on a circular embankment thrown up for the purpose. The intervening spaces were then, in many instances, filled in with small stones, so as to form a compact kind of wall, as shown in the next engraving. This mode of construction

Fig. 70.

was adopted for encircling grave-mounds, and in the forming of hut dwellings, etc. It will easily be seen that in course of time the loose walled parts would be thrown down and disappear, while the uprights, being firmly fixed in the ground, would remain, and would thus form the stone circles as now seen, and as commonly called "Druidical circles." In some instances, as in the case of the circle enclosing a perfect stone cist, covered by a mound, at Sancreed, shown on fig. 71, the upright stones touched each other, and thus formed a remarkably fine enclosure. This circle is about fifteen feet in diameter. Another

Fig. 71.   Fig. 72.

variety is shown in fig. 72. This is a double circle, or rather two circles, one within the other, and about two feet apart, surrounding a stone cist. The stones in this example nearly touch each other. A somewhat similar one, but with the circles farther apart from each other, exists in the Isle of Man, and is shown on the ground plan (fig. 73).

Fig. 73.

The mound in this instance, probably, rose from the inner circle only, and covered the central cist. In several instances the interment was not in the centre of the circle, but was made in different situations within its area. For instance, in the next example (fig. 74), from Trewavas

Head, the cist is near to the circle of stones. The outer diameter of the mound is thirty-five feet, the diameter of

Fig. 74.

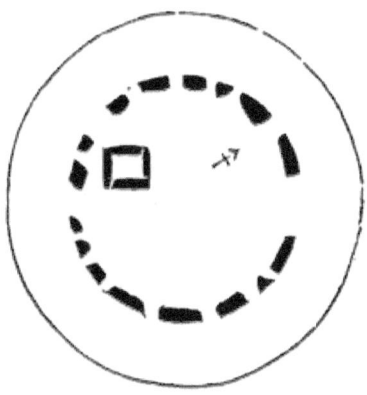

the circle of stones being nineteen feet six inches. Other examples, similar in arrangement, might be adduced.

Fig. 75.

Fig. 75 shows a totally different construction. In this instance the circle is composed of a number of stone cists, or sepulchral chambers, pretty close together, end to end. This curious example, of which a somewhat analogous one exists in the Channel Islands (see fig. 76), is on Mule Hill,

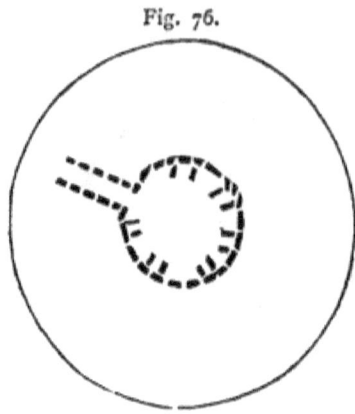

Fig. 76.

in the Isle of Man. Fig. 77 shows the remains of a stone circle surrounding the larger of a pair of "twin-barrows,"

Fig. 77.

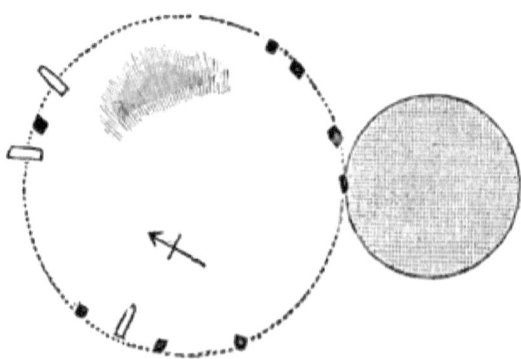

of which some of the stones have now disappeared. The circle is about seventy feet in diameter, and the stones vary from six to eight feet in height. Fig. 78 is the plan of another "twin-barrow," so called, the circle

Fig. 78.

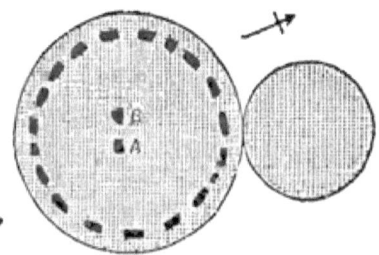

in the larger being about thirty-five, and the smaller twenty-four, feet in diameter. In the centre, at A B, are the remains of a stone cist, or chamber. "The mounds were both cairns of loose stones. Remains of other barrows, similarly formed, occur in the vicinity. There were two within a few hundred yards of the 'twin-barrow' last described, the greater portions of which have recently been taken away to build a neighbouring hedge, but of which I found enough to show how they were built. First, there was an enclosing circle of stones, some placed upright, some longitudinally (fig. 79), the intention being

Fig. 79.

simply to make an enclosing fence; within this the grave was constructed; then small stones heaped over the whole, the cairn extending, by about six feet, outside the built circle." The more perfect of the "twin-barrows" also had the cairn extending beyond the circle.

Some larger circles, such, for instance, as the Bosawen-ûn

circle, eighty feet in diameter (fig. 80), the Aber circle (fig. 81), and others, it is supposed, may have been formed

Fig. 80.

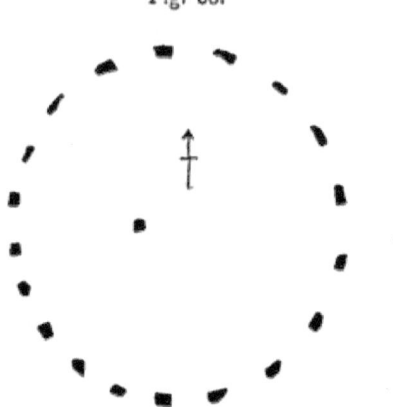

around a group of interments, instead of single interments, as in many of the others. In some instances a single stone

Fig. 81.

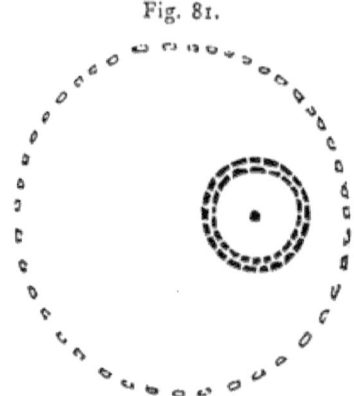

was placed to mark the place of interment. Three such exist in the barrow at Berriew (fig. 82). A large circle (fig. 83), twenty-seven yards in diameter, on Penmaenmaur, was constructed of several uprights, connected by smaller masonry. Here the interments were apparently

Fig. 82.

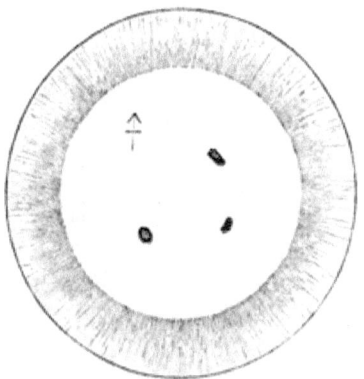

made beside the pillars. Against the inner side of the tallest pillar A, on the eastern part, were the remains of a small stone cist; while against the pillar B, facing it on the

Fig. 83.

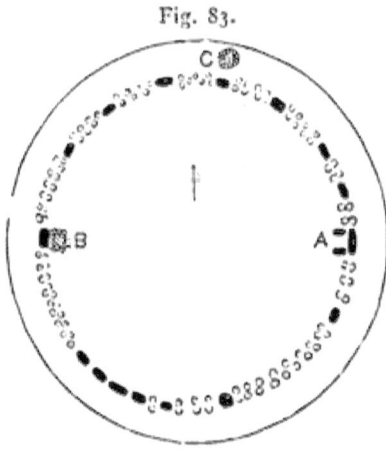

opposite side, was heaped up a small cairn. The whole is surrounded by a ditch, within which, at C, is another small cairn.

Arbor-Low, in the High Peak of Derbyshire, to which allusion has been made, is represented in fig. 84. No sepulchral remains have been discovered within the circle,

Fig. 84.

but barrows of great extent, which have yielded important remains on being excavated, are closely connected with it. It is, however, probable, that interments have existed, and been removed in past ages.

CHAPTER V.

Ancient British or Celtic Period—Pottery—Mode of manufacture—Arrangement in classes—Cinerary or Sepulchral Urns—Food Vessels—Drinking Cups—Incense Cups—Probably Sepulchral Urns for Infants—Other examples of Pottery.

HAVING spoken of the principles of construction of the grave-mounds of the Celtic period, and described the various modes of interment which they exhibit, I now proceed to speak of the objects found in them. Before doing so, however, it is necessary to say, that in the course of examination of barrows of this period it not unfrequently happens that the spot where the funeral pyre has been lit can very clearly be perceived. In these instances the ground beneath is generally found to be burned to some considerable depth; sometimes, indeed, it is burned to a fine red colour, and approaches somewhat to brick. Where it was intended that the remains should be collected together, and placed in an urn for interment, I apprehend, from careful examination, that the urn, being formed of clay —most probably, judging from the delicacy of touch, and from the impress of fingers which occasionally remains, by the females of the tribes—and ornamented according to the taste of the manipulator, was placed in the funeral fire, and there baked, while the body of the deceased was being consumed. The remains of the calcined bones, the flints, etc., were then gathered up together, and placed in the urn; over which the mound was next raised. When it was not intended to use an urn, then the remains were collected together, piled up in a small heap, or occasionally enclosed in a skin or cloth, and covered to some little thickness with earth, and occasionally with small stones. Another fire was then lit on the top of this small mound, which had the effect of baking the earth, and enclosing the remains of calcined bones, etc., in a kind of crust, resembling in colour and hardness a partly baked brick. Over this, as usual, the mound was afterwards raised.

The most important feature in the construction of the grave-mounds of the Celtic period is, perhaps, the pottery, and to this, therefore, the present chapter will be devoted. The pottery of this period may be safely arranged in four classes;* viz., 1. *Sepulchral Urns*, which have contained, or been inverted over, calcined human bones. 2. *Food Vessels* (so called), which are supposed to have contained an offering of food, and which are more usually found with unburnt bodies than along with interments by cremation. 3. *Drinking Cups*, which are usually ornamented. 4. *Incense Cups* (erroneously so called for want of more knowledge of their use), which are very small vessels, found only with burnt bones, and usually containing them, in the large cinerary urns.

The pottery was, without doubt, made on, or near to, the spot where found. It was, there is every probability, the handiwork of the females of the tribe, and occasionally exhibits no little elegance of form, and no small degree of ornamentation. The urns, of whatever kind they may be, are formed of the coarse common clay of the district where made, occasionally mixed with small pebbles and gravel; they are entirely wrought by hand, without the assistance of the wheel, and are, the larger vessels especially, extremely thick.

From their imperfect firing, the vessels of this period are

---

* For articles upon this subject see the "Reliquary, Quarterly Archæological Journal and Review," vol. ii., pages 61 to 70; and Mr. Bateman's "Ten Years' Diggings," page 279.

usually called "sun-baked" or "sun-dried," but this is a grave error, as any one conversant with examples cannot fail, on careful examination, to see. If the vessels were "sun-baked" only, their burial in the earth—in the tumuli wherein, some two thousand years ago, they were deposited, and where they have all that time remained—would soon soften them, and they would, ages ago, have returned to their old clayey consistency. As it is, the urns have remained of their original form, and although, from imperfect baking, they are sometimes found partially softened, they still retain their form, and soon regain their original hardness. They bear abundant evidence of the action of fire, and are, indeed, sometimes sufficiently burned for the clay to have attained a red colour—a result which no "sun-baking" could produce. They are mostly of an earthy brown colour outside, and almost black in fracture, and many of the cinerary urns bear internal and unmistakable evidence of having been filled with the burnt bones and ashes of the deceased, while those ashes were of a glowing and intense heat. They were, most probably, fashioned by the females of the tribe, on the death of their relative, from the clay to be found nearest to the spot, and baked on or by the funeral pyre. The glowing ashes and bones were then, as I have already stated, collected together, and placed in the urn, and the flint implements, and occasionally other relics belonging to the deceased, deposited along with them.

The *Cinerary*, or *Sepulchral*, *Urns* vary very considerably in size, in form, in ornamentation, and in material—the latter, naturally, depending on the locality where the urns were made; and, as a general rule, they differ also in the different tribes. Those which are supposed to be the most ancient, from the fact of their frequently containing flint instruments along with the calcined bones, are of large size, ranging from nine or ten, to sixteen or eighteen inches in height. Those which are considered to belong to a somewhat later period, when cremation had again become general, are of a smaller size, and of a somewhat finer texture. With them objects of flint are rarely found, but articles of bronze are occasionally discovered. The general form of the cinerary urns will be best understood from the annexed engravings.

The principal characteristic of the cinerary urns found in Derbyshire and Staffordshire, and in some other districts, is a deep overlapping border or rim, and their ornamentation, always produced by indenting or pressing twisted thongs into the soft clay, or by simple incisions, or by indentations produced by simple means, as will be more particularly named later on, is frequently very elaborate. It usually consists of diagonal lines (see fig. 85) arranged in a variety

Fig. 85.

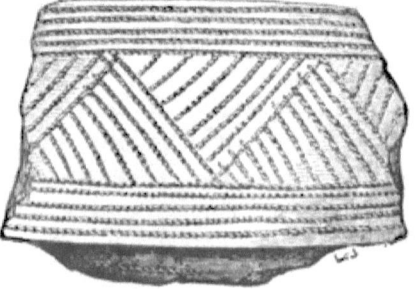

of ways, or of herring-bone or zigzag lines, or of reticulations, or of rows of punctures, etc., etc. This ornamentation is usually confined to the upper portion of the urn, including the over-lapping rim and the neck; and in many instances the upper edge and the inside of the rim were in like manner ornamented. Some of the more usual forms are the following.

Fig 86, from a barrow at Monsal Dale, was found along with many other interesting relics. It is twelve inches in height, and has a deep overlapping border. When found, it was inverted over a deposit of calcined bones placed on some rough stones on the natural surface, and having among them a calcined bone pin. Near it was a large mass of limestone, and a celt-shaped instrument five inches long,

Fig. 86.

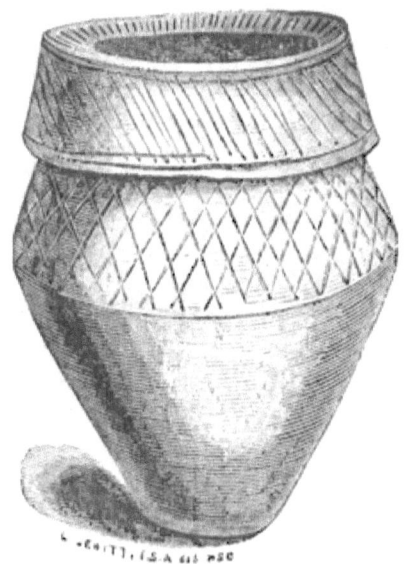

with a cutting edge, formed from the lower jaw of some animal. Another excellent example is exhibited in the urn from Ballidon Moor (fig. 87). It is eleven and a half inches in height, and measured nine inches in diameter at the mouth. It is ornamented by patterns impressed in the soft clay from a twisted thong. It contained burnt bones; amongst them were a portion of an animal's jaw, a fine bone pin, four inches long, rats' bones, a fragment of pottery, and a flint arrow-head. The presence of partially burnt human bones in the sand, the discolouration of the latter, and the occurrence of calcined rats' bones in the urns, demonstrated the fact of the corpse having been consumed

Fig. 87.

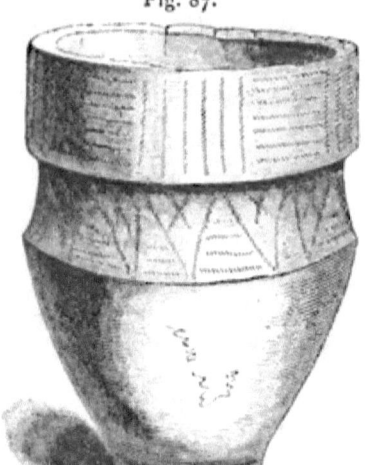

upon the spot. The following engraving (fig. 88), exhi-

Fig. 88.

biting a section of the barrow, will show the position of the urn when found, and also of the other interments which it contained.*

Fig. 89.

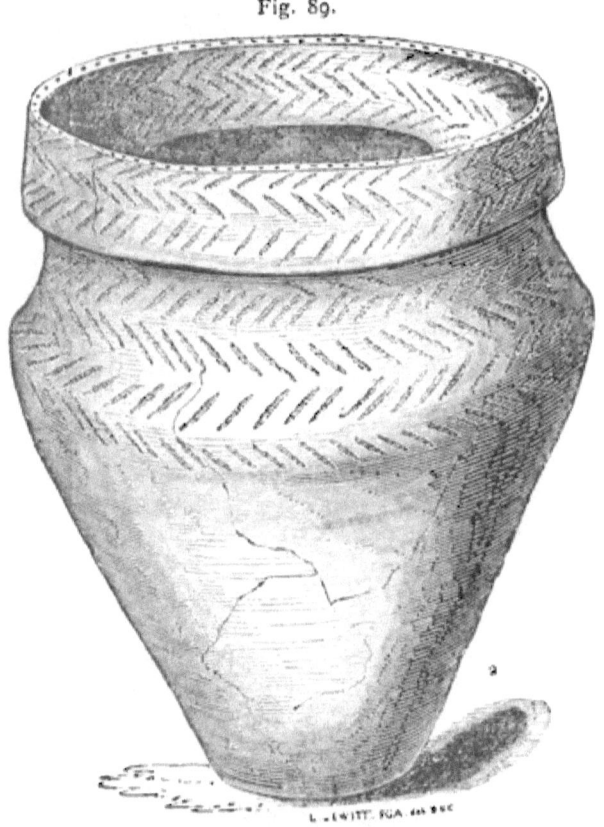

Fig. 89, from Trentham, Staffordshire, is a remarkably

* This barrow has been admirably described in that magnificent work, "CRANIA BRITANNICA,"—a work which every ethnologist and antiquary ought to possess, and which contains far more information than any other book extant. The following extract from the work explains the section:—
"Above this cist a cairn of fragments of sandstone had been raised most likely before interments by cremation were practised on the spot. The dark horizontal line of our woodcut indicates the situation of a stratum of burnt earth traversing the barrow at this height. Funereal

fine urn of the same character as the preceding examples,

Fig. 90.

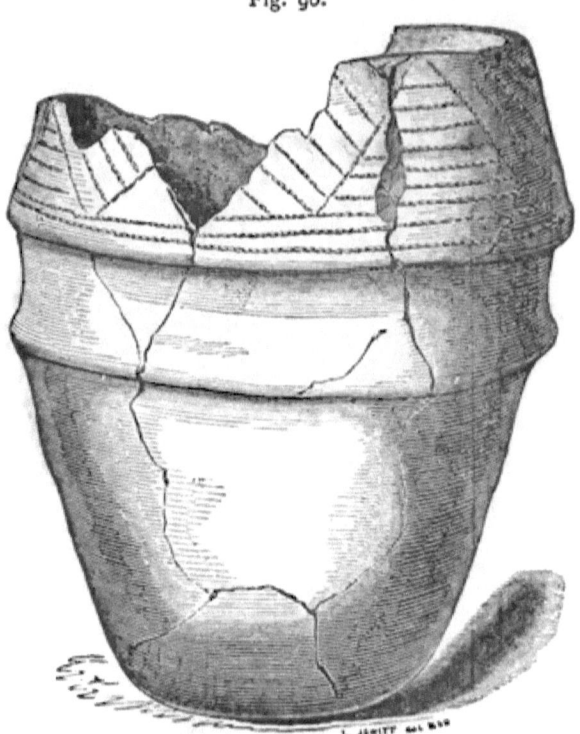

and fig. 90, from Darwen, Lancashire, has a central band

rites, by incineration, had evidently been celebrated on this surface; which was scattered over with a thin layer of wood-charcoal. In the centre of the barrow, and resting upon this carbonaceous deposit, stood a fine urn of dark British pottery, 11 inches high, and 9 at its greatest diameter at the top; not in the more commonly inverted, but in an upright position. It is ornamented in the usual style of lineal impressions, most probably made by a twisted thong of untanned leather, with rows of lines, alternately upright and horizontal, around the upper division; and in the middle the lines are varied into the zigzag, having distinct *crosses* and other impressions in the intervals. It contained calcined bones in a clean state, and mingled with them a portion of the jaw of some animal; bones of the water-vole (*Arvicola amphibius*, Desmar.), so common in the Derbyshire barrows; a bone pin, 4 inches in length,

as well as the overlapping rim. Figs. 91 and 92 are of totally different form; their ornamentation consisting of

Fig. 91.

Fig. 92.

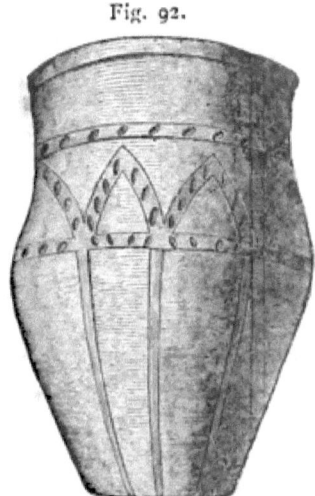

incised lines and impressed thumb marks, etc.* They are

and finely pointed; and a flint arrow-head; all calcined. The urn was closed by a large flat stone, the two ends of which rested upon side walls, so as to protect the deposit, and secure it from superincumbent pressure. Did this urn contain the inconsiderable yet sacred remains of one whose devotion in life the distinguished dead below had oft experienced—one who held life itself subordinate to his fate? The fearful conjecture seems not by any means improbable.

"Interred in the soil above this portion of the barrow, and lying amongst loose stones, the remains of four other skeletons occurred, placed in the primitive flexed position. One of these had apparently been disturbed at no long period subsequent to interment, and the bones laid in order before they had become decayed—a practice adopted by some uncivilized people in more modern times.

"This barrow of the British period presents unquestionable evidences of very primeval times, and contained the relics of a true aboriginal inhabitant of these islands, piously laid in his last resting-place with great care, but in all rude simplicity. It is rich in instruction, and marked by precise phases of information. It shows almost certainly the contemporaneous adoption of inhumation and cremation—the latter, perhaps, yielding to the first a short precedency; or possibly, in this instance, a rite of the nature of a "Suttee," and subordinate to the former.

* Warne's "Celtic Tumuli of Dorsetshire."

from Dorsetshire. The next example, from Darley Dale

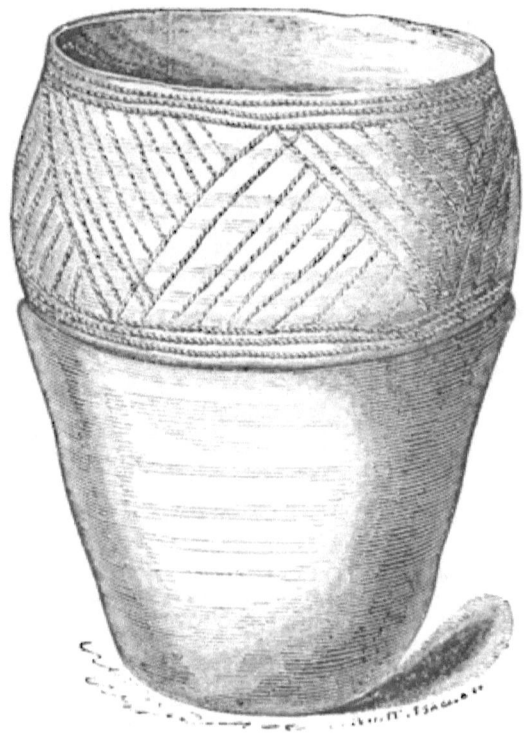

(fig. 93), is of a different type, as are also figs. 94, from Stone, Staffordshire, and 95 and 96, from Cleatham, in Lincolnshire. Other forms, again, are shown on fig. 97, from the Calais Wold-barrow, Yorkshire, discovered by Mr. Mortimer. It is eleven inches in height, and is ornamented with a number of small semi-punctures. A very fine urn was discovered by the Rev. Canon Greenwell in a barrow on Sutton Brow, near Thirsk, in the same county. It is sixteen inches in height, and eighteen in width, and is ornamented

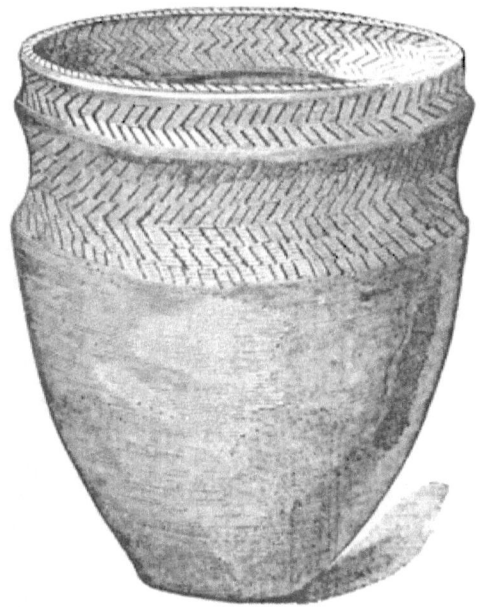

Fig. 94.

Fig. 95.   Fig. 96.

with lines produced by an impressed cord or thong, and by semi-punctures or indentations. The next example (fig. 98) is from Darley Dale, and is, as will be seen from the engrav-

Fig. 97.

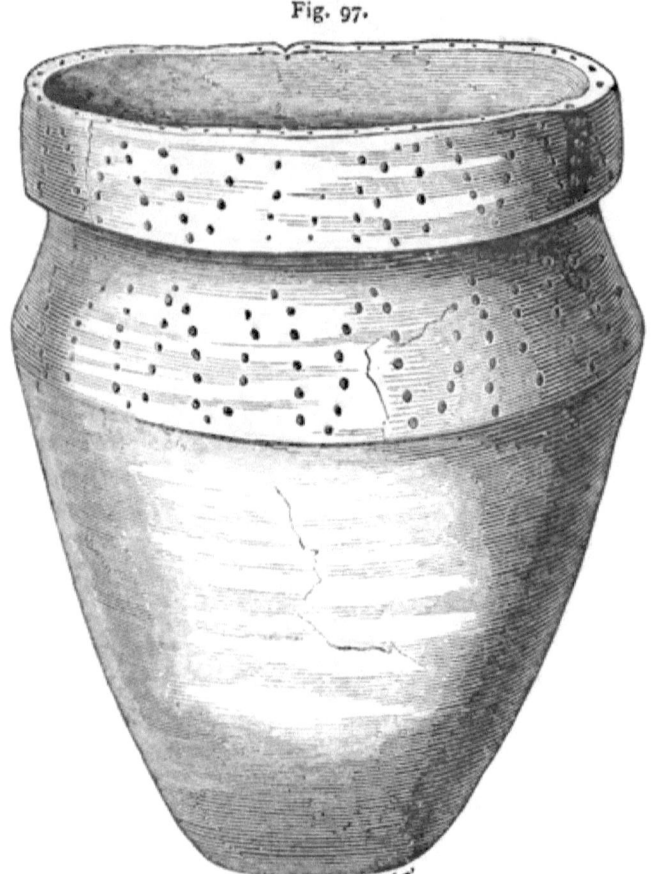

ing, of a very different character from the other examples given. Around the upper portion are encircling lines, between which is the usual zigzag ornament. Around the central band, too, are encircling lines, between which are a series of vertical zigzag lines. The whole of the ornamentation has been produced by pressing twisted thongs into the pliant clay—some, however, being of much tighter twist than others. Inside, the rim is ornamented with encircling

Fig. 98.

and diagonal lines. It has on its centre band four projecting handles or loops, which are pierced, as shown in the engraving. Another form, with small loops on its sides, is shown on fig. 99, which was found in one of the Cornish barrows, as was also fig. 100, which appears to have a kind of ear, or semi-handle, at its sides.

The *Food Vessels*, the next division, vary considerably,

Fig. 99.

in form, in size, and in ornamentation, from the very

Fig. 100.

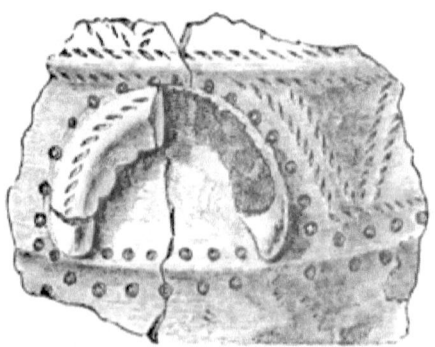

rudest to the most elegant and elaborate. These vessels are

generally wide at the mouth, and taper gradually downwards from the central band. They are found both where the interments have been by inhumation and by cremation, but much more frequently with the former. In these instances they are more usually placed near the head of the skeleton than in other positions, although they are occasionally found placed otherwise. Their average size is from four to six inches in height, and the ornamentation is produced in the same manner as has already been spoken of in reference to the cinerary urns, viz., by impressing twisted thongs or cords into the soft clay, by punctures, and by indentations produced in a variety of ways.

The "food vessels," like the cinerary urns, have evidently been made from the clay of the district where the interment has taken place, and they have been "fired" to about an equal degree of hardness with them.

Their general form will be best understood from the following examples, chosen from different districts.

Fig. 101.

The first example (fig. 99), from Trentham, in Staffordshire, is, it will be seen, of very rude form and make, and its ornamentation of simple character. Fig. 102, from

Fig. 102.

Fimber, in Yorkshire (5¾ inches in depth and 6¼ inches

Fig. 103.

wide at the mouth), is of a more usual form, and is more

advanced in point of ornamentation. Fig. 103 is from Hitter Hill, Derbyshire, as is also fig. 104. They were found in the interments shown on figs. 10, 11, and 12.

Fig. 104.

The first of these urns is four and three quarter inches in height, and five and a half inches in diameter at the top. It is richly ornamented with the usual diagonal and herring-bone lines, formed by twisted thongs impressed into the soft clay, in its upper part. Around the body of the urn itself, however, is a pattern of lozenge form, very unusual on vessels of this period. The second urn is five and a quarter inches in height, and six and a quarter inches in diameter at the top. It is very richly ornamented with the characteristic patterns found on the Celtic urns of this district, and is one of the finest and most elaborately ornamented which has been exhumed.

On Wykeham Moor, Yorkshire, some urns of a different form, wide at the mouth, were discovered by that hardworking antiquary, the Rev. Canon Greenwell.

Fig. 105 is from Monsal Dale, Derbyshire, and fig. 106,

from Fimber, Yorkshire, was found along with fig. 107.

Fig. 105.

These, as will be at once seen, are of a different character

Fig. 106.

from the preceding examples, in so far that on four sides

they have in the central sunk band a kind of handle or raised stud, which in some instances is pierced in the same manner as the cinerary urn (figs. 98 and 99). They are among the most elaborate, in point of ornamentation, of any of these interesting vessels. Other forms, besides those indicated, are occasionally found.

The *Drinking Cups* are the most highly and elaborately ornamented of any of the varieties of Celtic fictile art found in barrows. They are found with the skeleton, and are usually placed behind the shoulder. In size they range from about six to nine inches in height. They are usually tall in form, contracted in the middle, globular in their lower half, and expanding at the mouth. Their ornamentation, always elaborate, usually covers the whole surface, and is composed of indented lines placed in a variety of ways, so as to form often intricate, but always beautiful, patterns, and by other indentations, etc. They are much more delicate in manipulation than the other varieties of urns.

Instances have been known in which a kind of incrustation has been very perceptible on their inner surface, thus showing that their use as vessels for holding liquor is certain, the incrustation being produced by the gradual drying up of the liquid with which they had been filled when placed with the dead body.

Fig. 107, from a barrow at Fimber, is an elegant and highly characteristic example of this kind of vessel. It stood close behind the shoulders of the skeleton of a strong-boned middle-aged man, which lay on its right side. The ornamentation is most elaborate and delicate, and it is, perhaps, one of the finest and best preserved examples in existence.

The next two engravings (figs. 108 and 109) show two excellent examples, the first from the Hay Top barrow and the second from a barrow at Grind-Low, of a slightly

7.

Fig. 108.

Fig. 109.

different form at the mouth. The next example (fig. 110), found in Derbyshire, is of different shape, and has the unusual feature of being ornamented in quite as elaborate a manner on its bottom as it is around its sides. The bottom

Fig. 110.

is shown on fig. 111. The ornamentation throughout is produced by the indentations of twisted thongs into the soft clay. Figs. 112 and 113 are of a different form and character; the first of these is from Roundway Hill, Wiltshire (see fig. 8 for interment with which this interesting

Fig. 111.

Fig. 112.

vessel was found), and the second from "Gospel Hillock," in Derbyshire. Others of a similar form have been found also in Yorkshire and other counties.

Those which have been engraved are, perhaps, the most

Fig. 113.

usual of the forms of the drinking cups, but other shapes are occasionally discovered.

The next division, the so-called "*Incense Cups*," a name which ought to be discarded, consists of diminutive vessels which, when found at all (which is seldom) are found *inside* the sepulchral urns, placed on, or among, the calcined bones, and frequently themselves also filled with burnt bones. They range from an inch and a half to about three inches in height, and are sometimes highly ornamented, and at others plain.

The examples I here introduce (figs. 114 to 125) will give a good general idea of these curious little vessels, which I believe have not been "incense cups," but small urns to receive the ashes of an infant, perhaps sacrificed at the

Fig. 114. Fig. 115.

Fig. 116.

Fig. 117. Fig. 118. Fig. 119.

Fig. 120. Fig. 121.

CELTIC POTTERY—HANDLED VESSELS.

Fig. 122.   Fig. 123.

Fig. 124.   Fig. 125.

death of its mother, so as to admit of being placed within the larger urn containing the remains of its parent. The contents of barrows give, as I have before stated, incontestible evidence of the practice of sacrificing not only horses, dogs, and oxen, but of human beings, at the graves of the Ancient Britons. Slaves were sacrificed at their masters' graves; and wives, there can be no doubt, were sacrificed and buried with their husbands, to accompany them in the invisible world upon which they were entering. It is reasonable, therefore, to infer that infants were occasionally sacrificed on the death of their mothers, in the belief that they would thus partake of her care in the strange land to which, by death, she was removed. Whether from sacrifice, or whether from natural causes, the mother and her infant may have died together, it is only reasonable to infer from the situation in which these "incense cups" are found (either placed on the top of a heap of burnt bones, or inside the sepulchral urn containing them), and from their usually containing small calcined bones, that they were receptacles for the ashes of the infant, to be buried along with those of its mother.

The form will be seen to vary from the simplest salt-cellar-like cup to the more elaborately rimmed and ornamented vase. Some are pierced with holes, as if for suspension, and one or two examples have handles at the side. The best examples of this kind are those shown on figs. 120, 124, and 125.

Among the most curious vessels of this period may pos-

Fig. 126.

sibly be reckoned the singular one here engraved (fig. 126), of which form only two examples have been discovered. They are much in shape like the drinking cups before engraved, but have the addition of a handle at the side, which gives them the character of mugs. One of these is in the Ely museum, and the other in the Bateman museum.

CHAPTER VI.

Ancient British or Celtic Period—Implements of Stone—Celts—Stone Hammers—Stone Hatchets, Mauls, etc.—Triturating Stones—Flint Implements—Classification of Flints—Jet Articles—Necklaces, Studs, etc.—Bone Instruments—Bronze Celts, Daggers, etc.—Gold Articles.

THE implements of stone found in the Celtic gravemounds, or in their immediate neighbourhood, consist of celts* or adzes, hammer-heads or axe-heads, mauls, etc., etc. They are of various materials—chert, shale, greenstone, syenite, basalt, porphyry, felstone, serpentine, sandstone, limestone, etc., etc., and of various degrees of finish and workmanship.

Stone celts of one form or other are the most common of all stone implements. In shape they are not inaptly described as being like the mussel shell. The lower, or cutting end is slightly convex, and rubbed down to a fine-shaped edge. As this cutting edge has become dulled or chipped by use it has been again and again rubbed down and sharpened, until, in many instances which have come under my notice, the celt has been shortened fully one-third or more of its original length. The forms of these instruments will be seen in the examples here following (fig. 127 and in the succeeding figures). Fig. 132 is, perhaps, the most usual of these forms. It is of the same type as the first example on the previous engraving. Another excellent example is given on the illustration (fig. 134). It is of chert, and has, as will be seen, straight sides instead of the

* *Celt*, from *Celtis*, a chisel.

usual curved ones. It is now 5½ inches long, but has probably originally been much longer, having been rubbed down in sharpening.

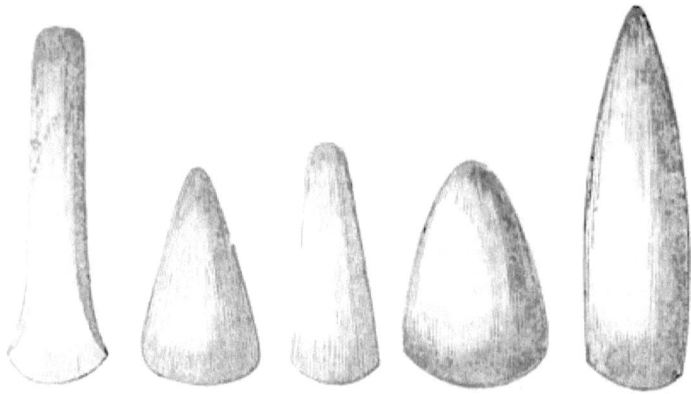

Fig. 127.   Fig. 128.   Fig. 129.   Fig. 130.   Fig. 131.

Stone hammers are occasionally found in grave-mounds. They vary much both in form and size, as will be best

Fig. 132.

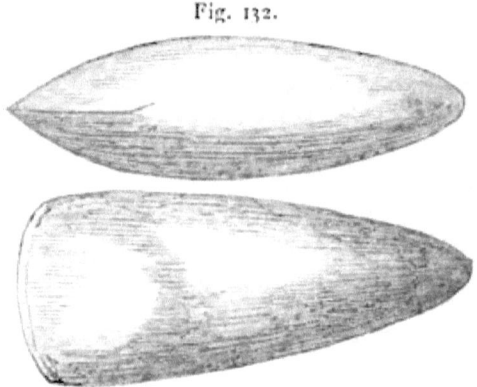

understood from the following engravings. Fig. 133 was found at Woolaton, and is remarkable for being hollow on its upper and lower surfaces, and ribbed or fluted along

Fig. 133.

its sides. It is eleven inches* long, four inches in width,

Fig. 134.

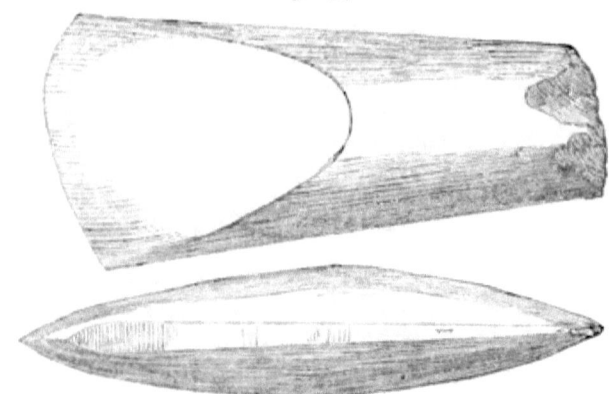

and three inches in thickness. Fig. 135, found at Winster,

Fig. 135.

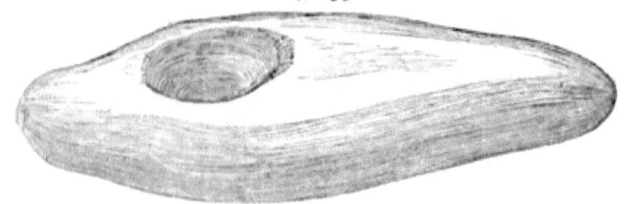

is thin, very taper, and of very different form. It is ten

Fig. 136. Fig. 137.

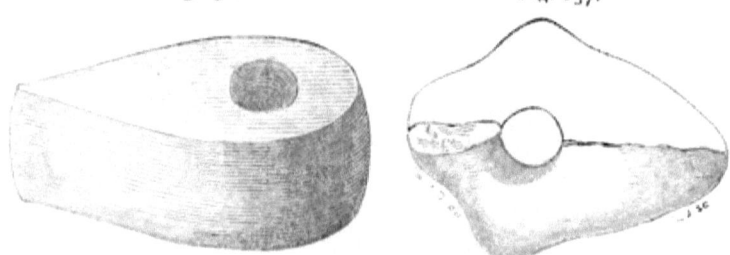

inches long. Other examples are shown in figs. 136 and

* This is one of the largest examples which have been found. It is

137. Occasionally they partake more of the hatchet shape.

Fig. 138.

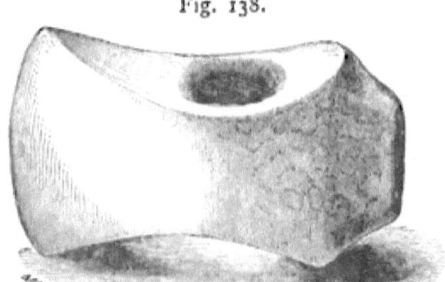

A good example is fig. 138, and others of still more elabo-

Fig. 139. Fig. 140. Fig. 141.

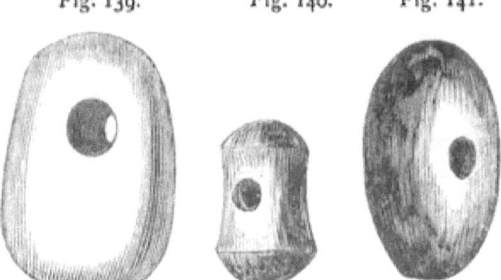

rate form have occasionally been discovered. Examples of

Fig. 142.

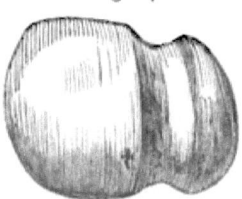

another variety, generally called mauls, which partake more of the common mallet form, are here given on figs. 139,

in my own collection, having been most kindly presented to me by the Hon. and Rev. C. Willoughby.

140, and 141. The first is from Horsley, Derbyshire, and the other two are from Ireland. A different variety (named punches or cutters) is shown on fig. 142, which was found at Mickleover.

Rough stones, which have probably been used for triturating purposes, for the grinding of corn, etc., are occasionally found. In the Derbyshire barrows, for instance, portions of rubbed stones, and also of rubbers, have now and then been discovered. Two triturating stones, belonging to a differ-

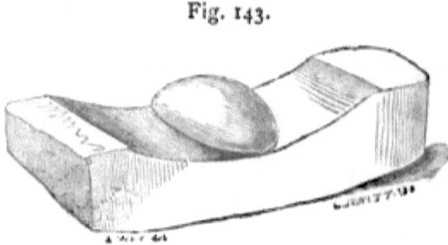

Fig. 143.

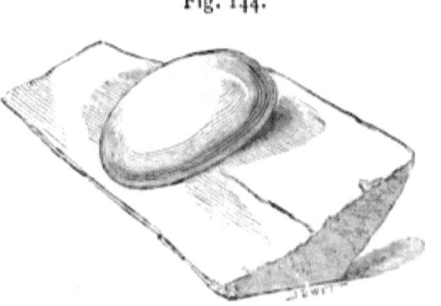

Fig. 144.

ent period, are given, for purposes of comparison, on figs. 143 and 144.* Whetstones, spindle-whorls, and other

* For a lengthened description and classification of the various forms of stone implements, the reader is referred to a new work, "The Ancient Stone Implements of Great Britain," by that able antiquary, Mr. John Evans, the author of the admirable volume on "Ancient British Coins," by which his name is so well known.

objects of stone, are also occasionally found. One of these spindle-whorls is shown on fig. 145.

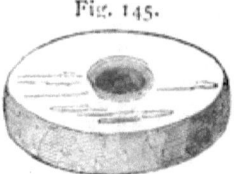

Fig. 145.

FLINTS, *i.e.*, various instruments formed of flint, are undoubtedly the most abundant of any relics of the Ancient Britons found in or about grave-mounds. They are extremely varied in form, and many of them are of the most exquisite workmanship—such, indeed, as would completely baffle the skill, great though that skill undoubtedly is, of "Flint Jack"* to copy. The arrangement, classification, and nomenclature of flints is at present so uncertain, and so mixed up with absurd theories, that it is difficult to know how to place them in a common-sense manner. All I shall attempt to do in my present work—which is intended to describe, generally, the relics to be found in the barrows of the period, and not to be a disquisition on flints alone—will be to give examples of some of the more usual forms which have from time to time been found, so as to facilitate comparisons with those of various districts and countries.

Of barbed arrow-heads, the examples here given will be sufficient to show the variety of forms and sizes which are usually found. The three first examples are from Green-Low, and are in the Bateman museum; the next three (figs. 149, 150, and 151) are also from Derbyshire examples in my own collection; fig. 152 is also from my own collection, but of a totally different form, approaching to the next example, fig. 153, which is in the museum of the Royal

* For a memoir, with portrait, of this remarkable character, and an account of his doings, see the *Reliquary*, vol. viii., p. 65, *et seq*.

116 GRAVE-MOUNDS AND THEIR CONTENTS.

Irish Academy. Fig. 149 will be noticed to be peculiarly elegant in form, and marvellously delicate in manufacture

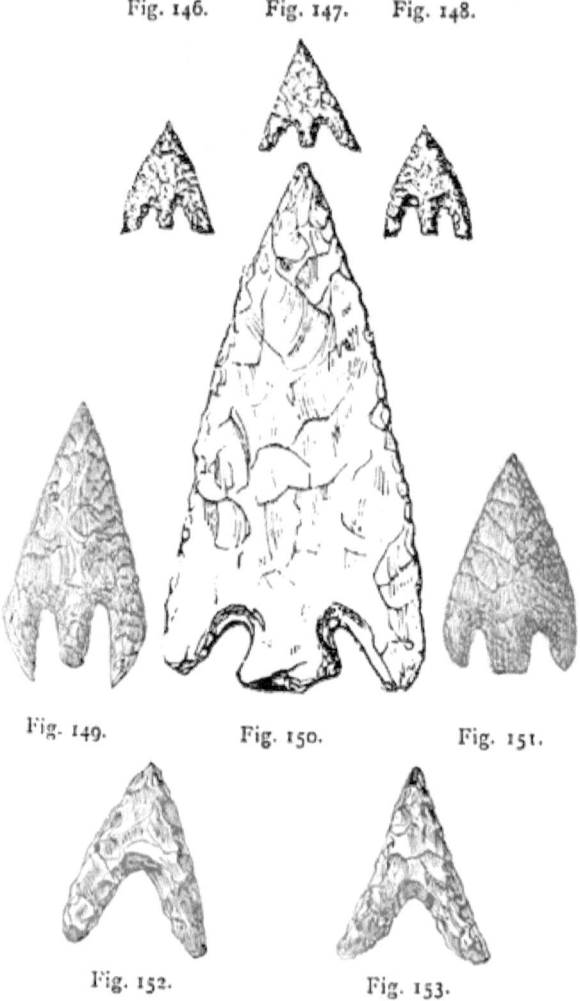

Fig. 146.   Fig. 147.   Fig. 148.

Fig. 149.   Fig. 150.   Fig. 151.

Fig. 152.   Fig. 153.

—the barbs being extremely sharp and clearly defined. It is engraved of its full size, as are most of the other examples. Fig. 150 measures two and five-eighths inches in length.

The dagger-blade variety is of what is usually called the "leaf-shaped" type, and is the prototype of the bronze dagger of a later period. The example here given (fig. 154)

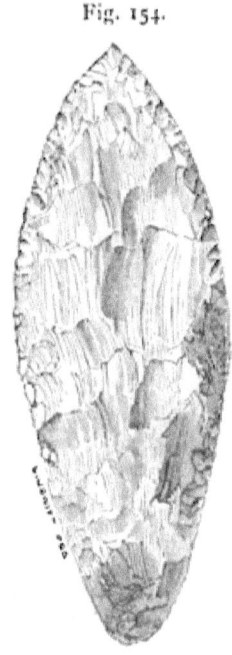

Fig. 154.

is from Green-Low, and is of remarkably fine form. Another, and of perhaps much finer form, is shown on the accompanying plate (fig. 155). It was found at Arbow-Low, in June, 1865, and is five and seven-eighths inches in length, and nearly two and a quarter inches in width in the centre. In its thickest part it is scarcely three-eighths of an inch in thickness, and is chopped and worked with the utmost nicety to a fine edge. It will be noticed that its sides, as they begin to diminish, are deeply serrated for fastening

Fig. 155.

with thongs to the haft or handle. It is engraved the exact size of the original.

The next illustrations exhibit a different variety of flints. They are arrow-heads of the leaf-shaped types, and exhibit four varieties. Figs. 156 and 157 are from Calais Wold, in Yorkshire; fig. 158 is from Gunthorpe, in Lincolnshire; and fig. 159, which is of remarkably elegant form, is from Ringham-Low, Derbyshire. They are engraved of their full size. This type of flint varies, it will be seen, from the acutely angled and sharply pointed shapes to those of a nicely rounded and egg-shaped form. Two other re-

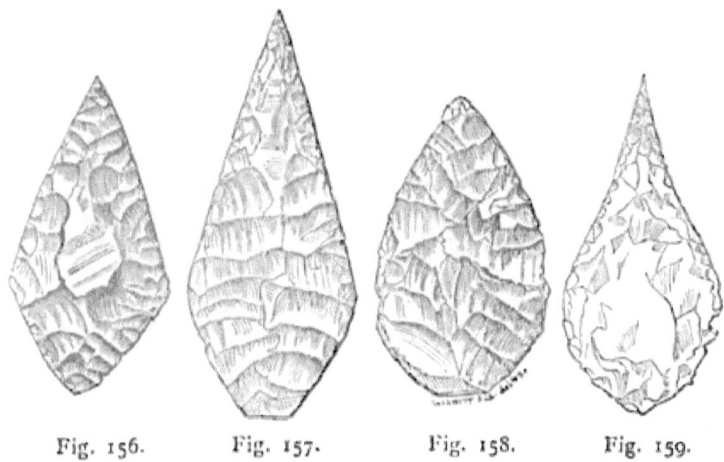

Fig. 156.   Fig. 157.   Fig. 158.   Fig. 159.

markable examples, possibly spear-heads, are here engraved, from the Calais Wold barrow, in Yorkshire (figs. 160 and 161). They are among the finest examples which have ever been found.

Another type, one not common in England, is shown on fig. 162. It is a fine example, and was found in Derbyshire. It is deeply serrated on the edges, and at its base is cut for tying with a thong. It is here engraved of its full size.

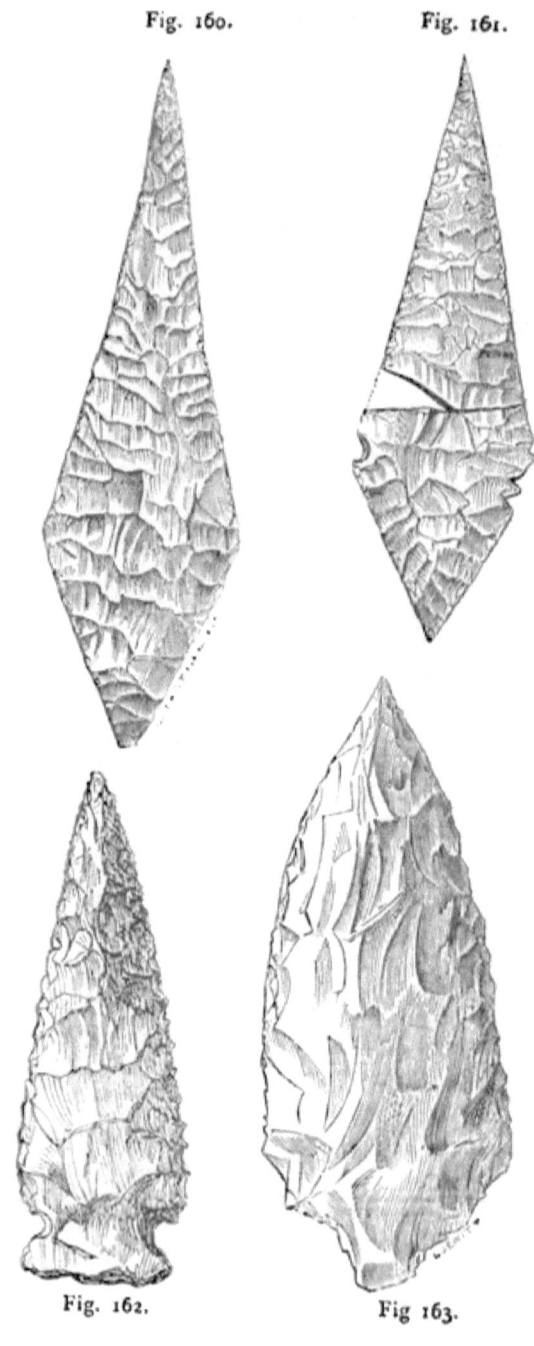

Fig. 160. Fig. 161.

Fig. 162. Fig 163.

Fig. 163 is a modification of this form, and is a good example of its kind. Figs. 164 and 165 are Derbyshire examples in my own collection, and are good specimens of another class of flint instruments not unfrequently found in grave-mounds and elsewhere.

Fig. 164.   Fig. 165.

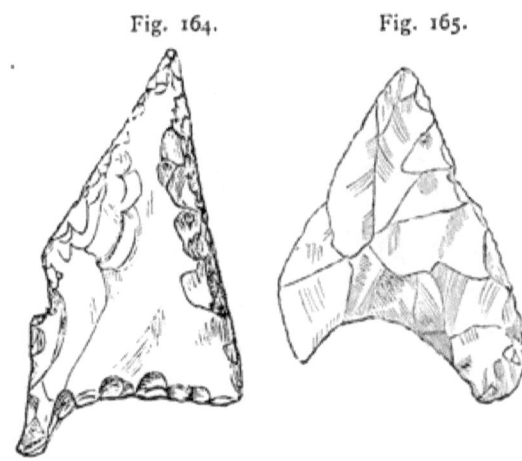

Another variety, again, and one which varies extremely, both in size and in form, is what, I suppose for want of a

Fig. 166.

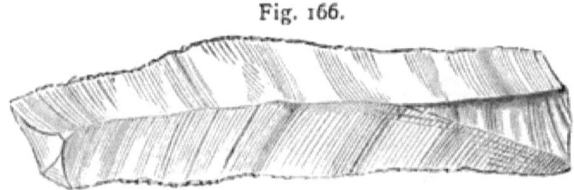

better name, is the kind usually called "scrapers," or "flint knives." One example (fig. 166) will be sufficient.

Another description, again, which appears more intended for throwing than for any other purpose, and which, with its sharp cutting edges, and the unerring aim of the Briton, must have been indeed a deadly weapon, is frequently found, and is shown on fig. 167. It is a simple circular lump of flint, an inch and a half or a couple of inches or more in diameter; flat on one side and chipped into a roundness on the other. These are often called "thumb flints."

Flakes of various sizes and forms constantly occur, and

Fig. 168.   Fig. 167.   Fig. 169.

are called by many absurd names. Small, delicately formed, and very beautiful flints, of an oviform or circular shape, are also found (fig. 168), as are a large number of

Fig. 170.

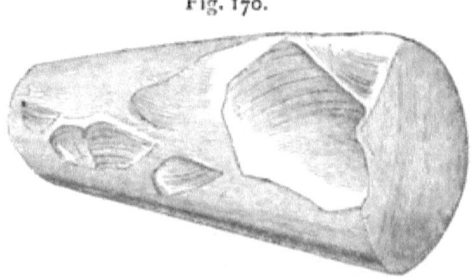

other forms besides those I have illustrated. These will, however, be sufficient for my present purpose, and will enable the reader to form a pretty correct and extended estimate of the number and variety of flints which the grave-mounds produce. Celts of flint are also occasionally found. An example here shown (fig. 170) was discovered in a very interesting barrow called "Gospel Hillock," at Cow Dale, near Buxton, by Captain Lukis. It measured four and a half inches in length.

Fig. 171.

In JET, the articles found consist of beads, rings, necklaces, studs, etc., and some of these are of the utmost beauty. A very elaborate example of necklace, found by

Mr. Bateman in the cist (fig. 28) on Middleton Moor, is here engraved (fig. 171). The beads of which it is composed lay about the neck of the skeleton. It was formed of variously shaped beads and other ornaments of jet and bone curiously ornamented. The various pieces of this elaborate necklace count 420 in number; 348 being thin laminæ, 54 of cylindrical form, and the remaining 18, conical studs and perforated plates, some of which are ornamented with punctures.

Another example (fig. 172), with elongated beads, and pierced ornaments of bone, is here given.

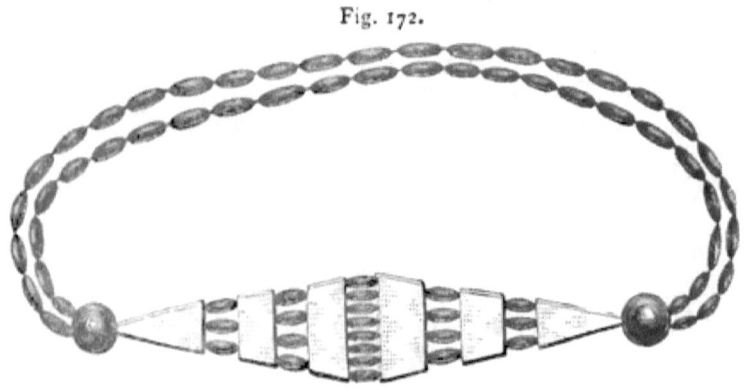

Fig. 172.

Another good example is engraved on the next page. It was found at Fimber, by Mr. Mortimer, and consists of 171 laminæ, or small jet discs (No. 2), and a triangular pendant, or centre, of jet (No. 3), an inch in length, and perforated in the middle.

Studs and pendants of jet are of various forms, and are perforated for suspension in a variety of ways. Fig. 174 shows a jet stud from Gospel Hillock. It is engraved of its full size, as is also the next example (fig. 175), from the Calais Wold barrow. These are very similar in form, and

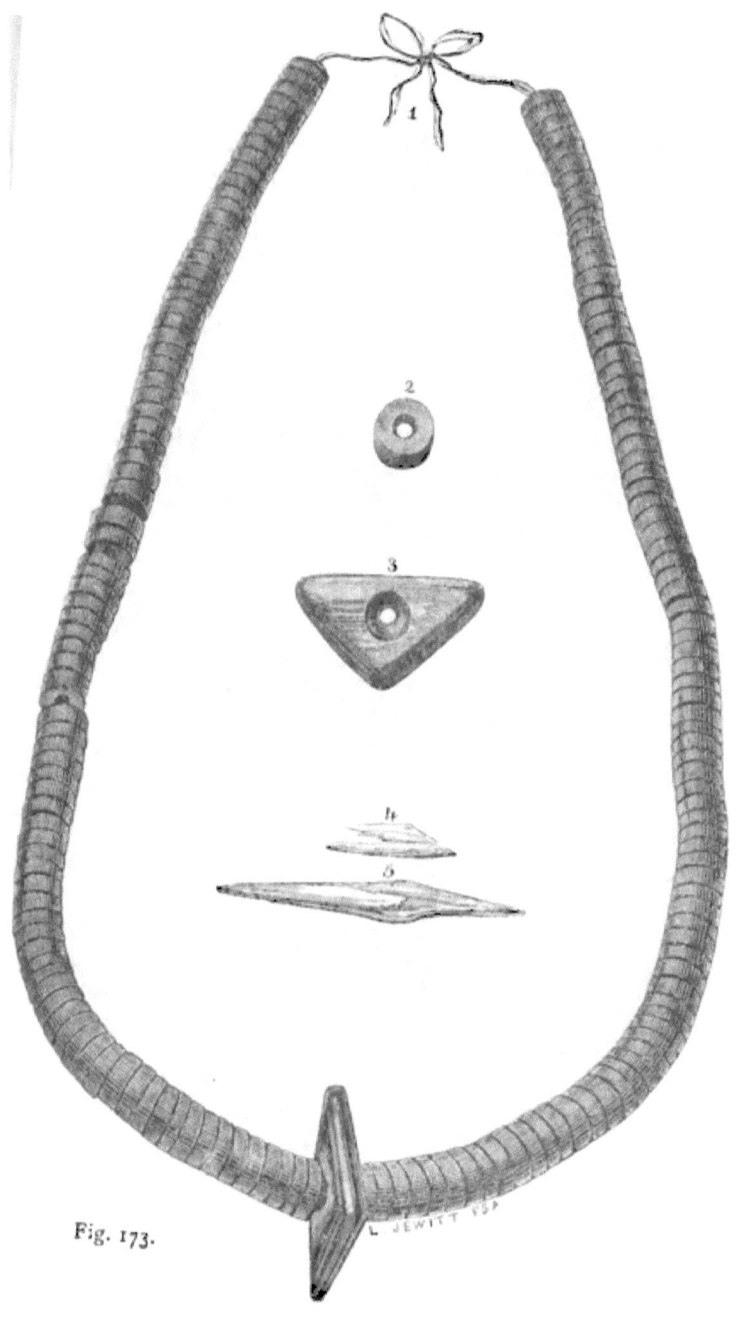

Fig. 173.

in their perforations. Another form, a ring pierced for suspension, is shown on fig. 176.

Implements of bone are frequently found, but in many instances their use is not easily determined. They consist chiefly of modelling tools (supposed to have been used in

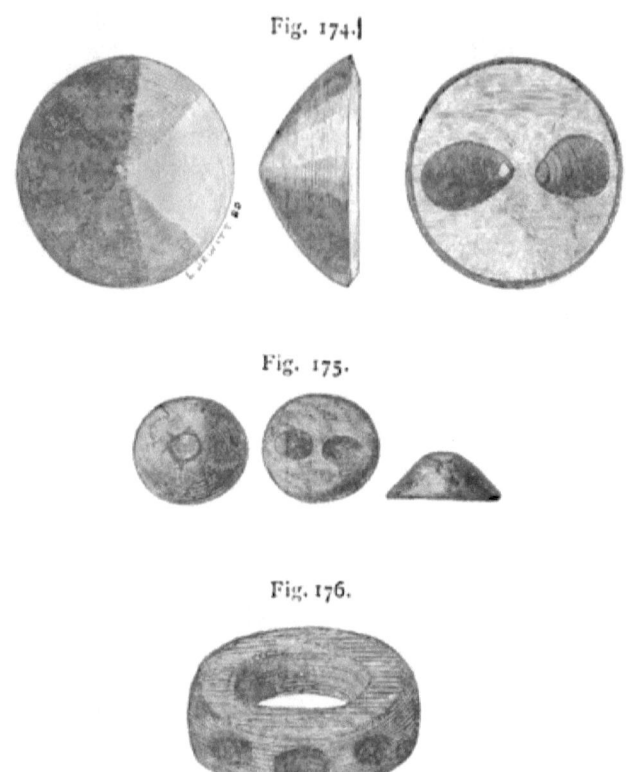

Fig. 174.

Fig. 175.

Fig. 176.

the manufacture of pottery), pins, mesh-rules, studs, pendants, and other personal ornaments; lance-heads, spearheads, whistles (?), hammers, and beads. Some of these are shown in figs. 177 to 182.

Fig. 177.

Fig. 178.

Fig. 179.

Fig. 180.

Fig. 181.

Fig 182.

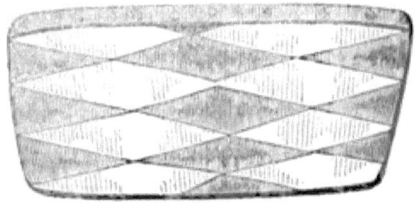

In BRONZE, the articles found are celts, daggers, awls, pins, etc. Celts are, however, but seldom met with in barrows, although frequently ploughed up in the course of agricultural operations. Palstaves and socketed celts, etc., are also occasionally picked up. The ordinary form of celt

Fig. 183.

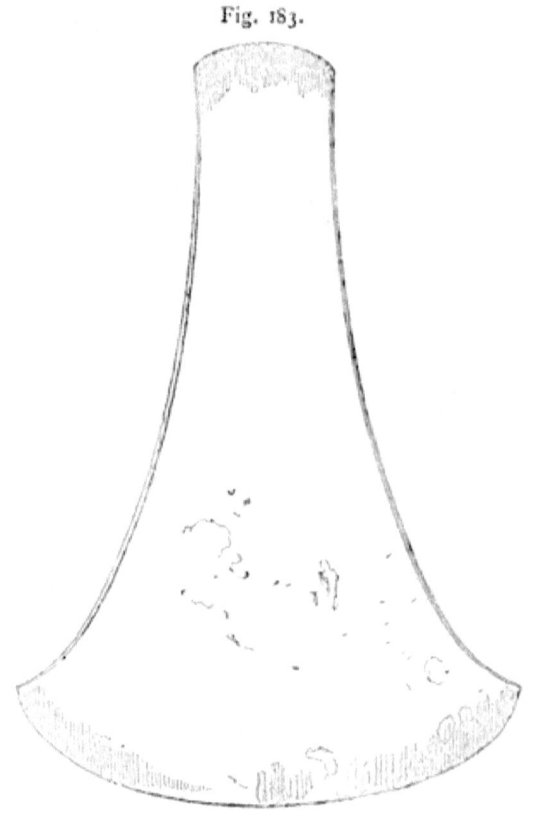

will be best understood by the engravings here given (figs. 183 and 185) from Irish examples, and by the next figure (187), from Moot-Low, near Dove Dale. One of these celts, of precisely similar form to fig. 187, found in a barrow

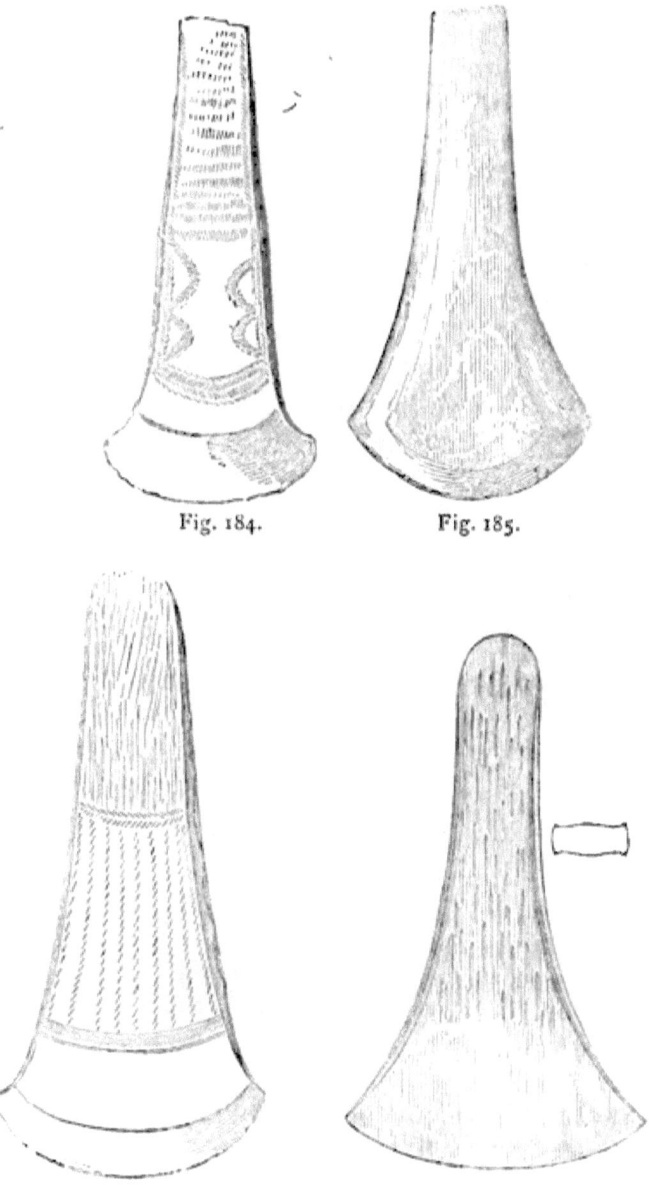

Fig. 184.   Fig. 185.

Fig. 186.   Fig. 187.

at Shuttlestone, has been the means of throwing considerable light on the mode of interment adopted. The barrow contained "the skeleton of a man in the prime of life and of fine proportions, apparently the sole occupant of the mound, who had been interred whilst enveloped in a skin of dark red colour, the hairy surface of which had left many traces both upon the surrounding earth and upon the verdigris or patina coating of a bronze axe-shaped celt and dagger, deposited with the skeleton. On the former weapon there are also beautifully distinct impressions of fern leaves, handfuls of which, in a compressed and half-decayed state, surrounded the bones from head to foot. From these leaves being discernible on one side of the celt only, whilst the other side presents traces of leather alone, it is certain that the leaves were placed first as a couch for the reception of the corpse, with its accompaniments, and after these had been deposited, were then further added in quantity sufficient to protect the body from the earth."* With the skeleton, besides the celt, were a fine bronze dagger, with two rivets for attachments to the handle, which had been of horn, the impression of the grain being quite distinctly perceptible; a small jet bead; and a circular flint. The celt had been, as was evident from the grain of wood still remaining, driven vertically, for about two inches of its length, into a wooden handle.

Other forms of celts are shown on the accompanying series of figures (184, 185, 186, and 188 to 195), and another excellent example is fig. 196, which has the loop (as also fig. 197) for attaching to the handle by means of a thong. A great many other varieties are also met with.

The bronze daggers which barrows have afforded vary in length from two and a half or three, to five and a half or six, inches, on the average; the larger ones being an

* "Ten Years' Diggings."

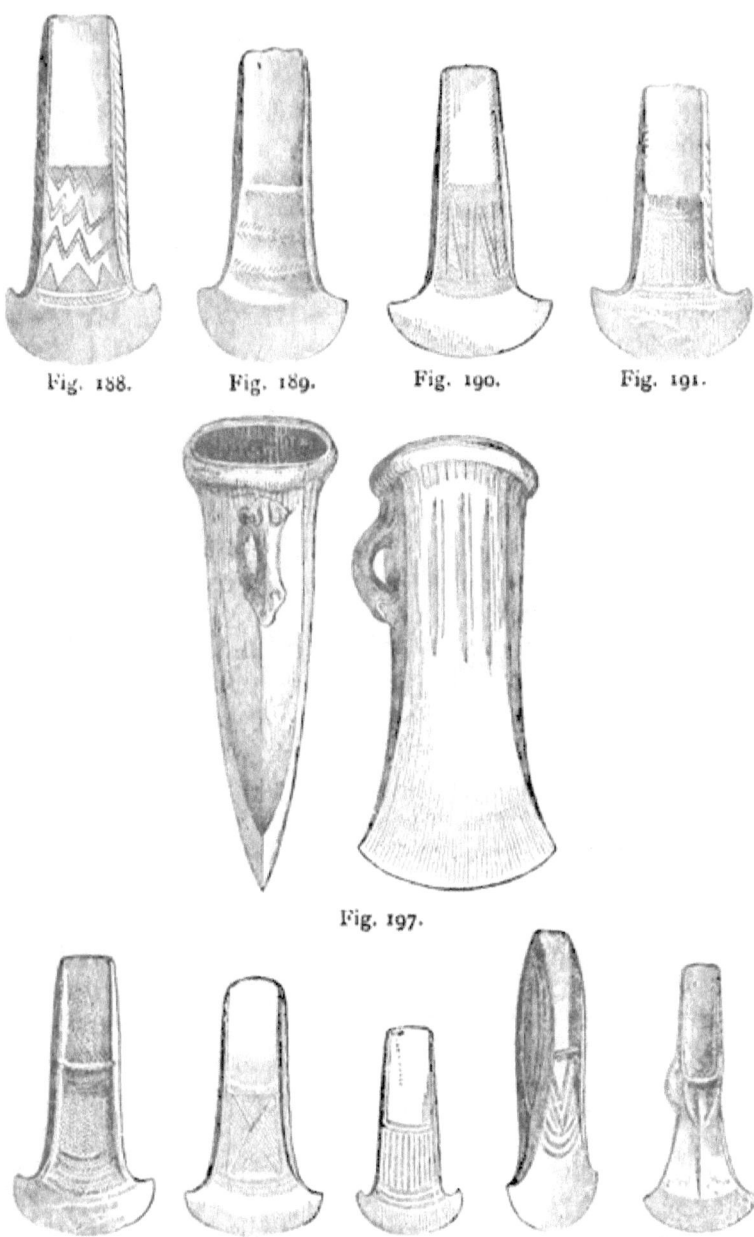

Fig. 188. Fig. 189. Fig. 190. Fig. 191.

Fig. 197.

Fig. 192. Fig. 193. Fig. 194. Fig. 195. Fig. 196.

inch and a half to three inches in breadth at their broadest part, where the handle has been attached, from whence they taper gradually down to the point. They are sometimes ribbed or fluted. In most instances the handle has been attached by three rivets; in some cases, however, as in fig. 198, only two have been used, and occasionally there

Fig. 198.

is evidence of the attachment being effected by thong or other ligature. The handles were of horn or wood, and were usually semi-lunar where attached to the blade; in one instance, however, the blade has a "tang" or "shank," which has fitted into the square-ended handle, to which it has been fastened by a single peg. The blades occasionally present incontestible evidence of long use, having been worn down by repeated sharpenings. In the instance of the dagger found at Stanshope, which had been fastened to the handle by a couple of rivets as well as by ligatures, evidence existed of its having been enclosed in a sheath of leather, and this example also presented the somewhat curious feature of impressions of maggots, which had probably made their way from the decaying body into the inside of the sheath, between it and the blade, and had there remained, and thus gradually become marked upon the corrugated surface of the bronze.

Articles of gold, and coins, are extremely rare as found in grave-mounds, although not unusually turned up in their neighbourhoods, and in places which have been inhabited by the pre-historic races. Simply for the purpose of showing the character of some of the Celtic coins, the following engravings are given.

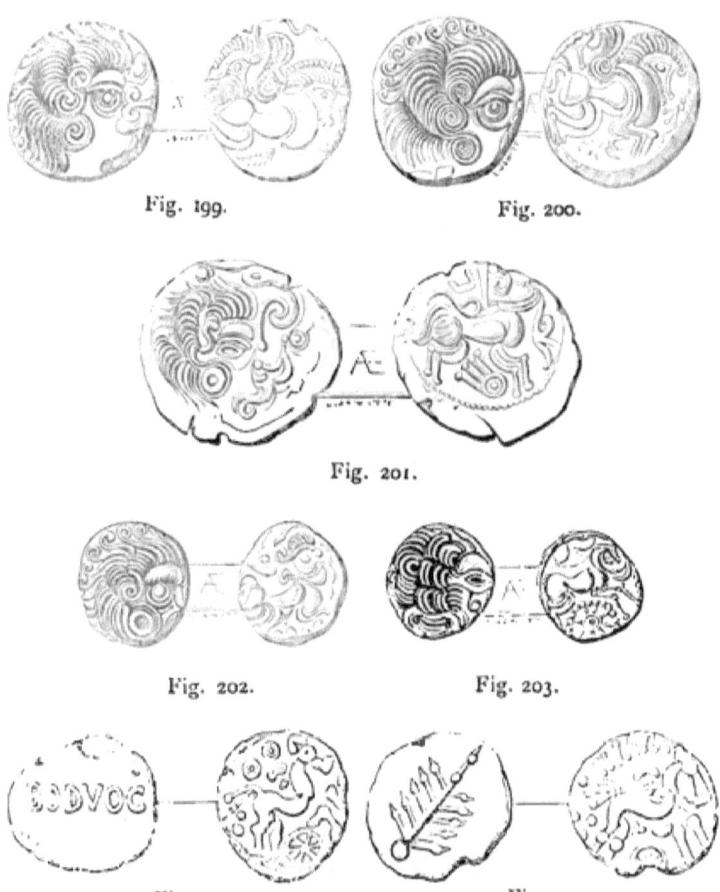

Fig. 199.　　　　　Fig. 200.

Fig. 201.

Fig. 202.　　　　　Fig. 203.

Fig. 204.　　　　　Fig. 205.

Of torques of gold, and other remains in that metal, I shall speak in a later chapter.

CHAPTER VII.

Romano-British Period—General characteristics—Modes of Burial—Customs attendant on Burial—Interments by cremation and by inhumation—Barrows—Tombs of Stone—Lead Coffins—Clay and Tile Coffins—Sepulchral Inscriptions, etc.

THE grave-mounds and burial-places of the Romano-British period are, naturally, in many districts, far more abundant than those of the preceding period, while, in others again, as in Derbyshire and Cornwall, and some other counties, they are far less common than the Celtic ones. In these counties the Roman was, it would seem, more of a "bird of passage" (as well as, to some extent, a "bird of prey") than a settler, and the consequence is, that no remains—or next to no remains—of villas or of settlements are found, and that where burial has taken place it has not unusually been in the same mound with those of an earlier period. The Ancient Briton raised the mounds over the remains of his own people; and his Roman subjugator, as occasion required, took possession of them, and therein laid his own dead. Thus the same barrow is sometimes found to contain, besides its primary Celtic interment, and others belonging to the same race, later deposits (nearer to the surface or to the side) of the Romano-British or of the Anglo-Saxon periods.

In other counties, where the Roman population made permanent settlements and built their towns and villas, regular cemeteries were formed for the burial of their dead, and to these mainly are we indebted for a knowledge of their customs and of their arts. The burials were, as in the previous period, both by inhumation and by cremation. The first appears to have been the most ancient practice of the Roman people, and it was not, as is stated, until the time of the dictator Sylla that burning of the dead was practised. From his time downward both of these usages were in vogue, according as the friends of the deceased preferred. So indiscriminately were these usages adopted in England that both are found in the same burial-places, and indeed (as in those of the Celtic period) in close proximity to each other.

The cemeteries attached to Roman towns were outside the walls, and usually by the road leading to the chief town—Londinium. In the country the owner of a villa had his burial-place in his own precincts.* Almost always, except when the interment was made in an earlier barrow, the dead were laid near to the living. In fact, the Roman seems, even when dead, to have still courted the proximity of the living, for he always by preference sought to establish his last home as near as possible to the most frequented road; and the inscriptions on his roadside tomb often contained appeals to the passers-by—in terms such as SISTE VIATOR (*stay, traveller*), or, TV QVISQVIS ES QVI TRANSIS (*thou, whoever thou art, who passest*)—to think on the departed. The epitaph on a Roman named Lollius, published by Grüter, concludes with the following words, intimating that he was placed by the roadside in order that the passer-by might say, " Farewell, Lollius ! "

HIC . PROPTER . VIAM . POSITVS
UT . DICANT . PRAETEREVNTES
LOLLI . VALE.

These examples will explain the position of the cemeteries of Uriconium and other Roman towns in Britain.

Mr. Wright, than whom no one is more able to speak

* *Reliquary* for October, 1861.

authoritatively on the matter, thus speaks of the burial customs and observances of the Romans in Britain; and as it is necessary, before speaking of the objects found with the sepulchral remains of the people, to give a sketch of the formalities attending their death and burial, his account will add considerable interest to their consideration.

"The last duty to the dying man was to close his eyes, which was usually performed by his children, or by his nearest relatives, who, after he had breathed his last, caused his body first to be washed with warm water, and afterwards to be anointed. Those who performed this last-mentioned office were called *pollinctores*. The corpse was afterwards dressed, and placed on a litter in the hall, with its feet to the entrance door, where it was to remain seven days. This ceremony was termed *collocatio*, and the object of it is said to have been to show that the deceased had died a natural death, and that he had not been murdered. In accordance with the popular superstition, a small piece of money was placed in the mouth, which it was supposed would be required to pay the boatman Charon for the passage over the river Styx. In the case of persons of substance, incense was burnt in the hall, which was often decked with branches of cypress, and a keeper was appointed, who did not quit the body until the funeral was completed. The public having been invited by proclamation to attend the funeral, the body was carried out on the seventh day, and borne in procession, attended by the relatives, friends, and whoever chose to attend, accompanied by musicians, and sometimes with dancers, mountebanks, and performers of various descriptions. With rich people, the images of their ancestors were carried in the procession, which always passed through the Forum on its way to the place of burial, and sometimes a friend mounted the rostrum, and pronounced a funeral oration. In earlier times the burial always took place by night, and was attended with persons carrying lamps or torches, but this practice seems to have been afterwards neglected; yet the lamps still continued to be carried in the procession. Women, who were called *præficæ*, were employed not only to howl their lamentations over the deceased, and chant his praises, like the Irish keeners, but to cry also; and their tears, it appears, were collected into small vessels of glass; and this circumstance is termed, in some of the inscriptions found on the Continent, being 'buried with tears' —*sepultus cum lacrymis;* and the tomb is spoken of as being 'full of tears'—TVMVL . LACRIM . PLEN.

"The next ceremony was that of burning the body. In the earlier ages of their history the Romans are said to have buried the bodies of their dead entire, without burning; and there seems to be no doubt that, at all events, the two practices, burning the body and cremation, existed at the same time; but the latter appears to have become gradually more fashionable, until few but paupers were buried otherwise. In the age of the Antonines the practice of cremation was finally abolished in Italy; but the imperial ordinances appear to have had but little effect in the distant provinces, where the two manners of burial continued to exist simultaneously. Both are accordingly found in the Roman cemeteries in Britain, in interments which were undoubtedly not those of Christians. Perhaps the practices varied in different parts of the island, according to the usages of the country from which the colonists derived their origin. It is a circumstance worthy of remark that, as far as discoveries yet go, no trace has been met with of burials in the Roman cemeteries of Uriconium, otherwise than by burning the dead.

"The funeral pile, *pyra*, was built of the most inflammable woods, to which pitch was added, and other things, which often rendered this part of the ceremony very expensive. An inscription, preserved by Grüter, speaks of some persons whose property was only sufficient to pay for the funeral pile and the pitch to burn their bodies—*nec ex eorum bonis plus inventum est quam quod sufficeret ad emendam pyram et picem quibus corpora cremarentur.* It had been ordered by a law of the Twelve Tables, that the funeral pile must be formed of timber which was rough and untouched by the axe, but this ru'e was perhaps not very closely adhered to in later times. When the body was laid on the pile, the latter was sprinkled with wine and other liquors, and incense and various unguents and odoriferous spices were thrown upon it. It was now, according to some accounts, that the *naulum*, or the coin for the payment of the passage over the Styx, was placed in the mouth of the corpse, and at the same time the eyes were opened. Fire was applied to the pile by the nearest relatives of the deceased, who, in doing this, turned their faces from it while it was burning; the relatives and friends often threw into the fire various objects, such as personal ornaments, and even favourite animals and birds. When the whole was reduced to ashes, these were sprinkled with wine (and sometimes with milk), accompanied with an invitation to the *manes*, or spirit of the deceased. The reader will call to mind the lines of Virgil (Æn. vi. 226):—

> ' Postquam collapsi cineres, et flamma quievit,
> Relliquias vino et bibulam lavere favillam,
> Ossaque lecta cado texit Corynæus aëno.'

"The next proceeding, indeed, was to collect what remained of the bones from the ashes, which was the duty of the mother of the deceased, or, if the parents were not living, of the children, and was followed by a new offering of tears. Some of the old writers speak of the difficulty of separating the remains of the burnt bones from the wood ashes, and we accordingly find them usually mixed together. When collected, the bones were deposited in an urn, which was made of various materials. The urn in Virgil was made of brass, or perhaps bronze. Instances are mentioned of silver, and even gold, being used for this purpose, as well as of marble; and those found in Britain are often of glass, but the more common material was earthenware. One of the performers in the ceremony, whose duty this was, then purified the attendants by sprinkling them thrice with water, with an olive branch (if that could be obtained), and the *præficæ* pronounced the word *Ilicet* (said to be a contraction of *Ire licet*, ' you may go'). Those who had attended the funeral, thrice addressed the word *Vale* (farewell) to the manes of the dead, and departed. A sumptuous supper was usually given after the funeral to the relatives and friends.

" In the case of people of better rank, the body was burnt on the ground which had been purchased for the sepulchre, but for the poorer people there was a public burning-place, which was called the *ustrina*, where the process was probably much less expensive, and whence the urn, with the remains (*relliquiæ*) of the deceased, was carried to be interred. The tombs of rich families were often large and even splendid edifices, with rooms inside, in the walls of which were small recesses, where the urns were placed. None of the buildings remain in any Roman cemetery in our island, but we can hardly doubt that such tombs did exist in the cemetery of Uriconium, and that they were scattered along the side of the Watling Street. At one place at Uriconium the foundations of a small building were met with, which appeared to have consisted of an oblong square, with a rectangular recess behind, but the western portion of it has been destroyed by the process of draining. When opened, ashes and fragments of an urn were found in the enclosed space, so that it is not improbable that this may have been a tomb with a room. An inscribed stone, which was found not far from this spot, bears evidence, in the appearance of its reverse side and in its form, of having been fixed against a wall, probably over a door." The urn was perhaps here interred beneath the floor of the room. In many cases the dead body was certainly burnt on the spot where it was to be buried. A square pit had been dug, on the floor of which the funeral pile had been laid. The fire had then been lit in the pit or grave, and the body consumed in its own grave. Remains of the timber of the funeral pile still remained in a pit of this kind at Uriconium, as it had sunk on the floor, the ends of which were unconsumed, and the earth underneath quite red from burning.

In most of the other interments in the cemetery of Uriconium, a small hole or pit appears to have been sunk in the ground, and the urn, which had no doubt been brought from the *ustrina*, was placed in it and covered up. These interments were not far distant from each other, and appear to have been placed in rows, nearly parallel to the road. Perhaps the ground may have been bought for this purpose in common, by associations of the townsmen, such as trade corporations, or it may have been set aside for burial purposes by the municipal authorities, and sold in small portions to individuals, as the practice now exists in modern cemeteries. The average depth at which the urns have been found is somewhat less than four feet, so that, allowing two feet for the accumulation of soil, the Romans seem to have dug pits about two feet deep for their reception.

Coins were, as has just been stated, buried with the dead, in conformity with a superstitious belief that they would expedite the passage of the soul across the lake in Hades. The magic power of money in all connections with human life originated this custom. In all worldly matters money then was, as it unfortunately now still is, the main, if not the only, sure passport to place and honour; and thus it was believed that the soul of the man who had not received the usual rites of burial, and in whose mouth no fee for the ferryman of the Stygian lake had been placed,* would wander hopelessly on its banks, while decent interment and a small brass coin would obviate any disagreeable inquiries that Charon might else be inclined to make as to the merits or claims of the applicant. Thus in the cinerary urns of the period of which I am speaking coins are very commonly found, and also in interments by inhumation a small coin has, in more than one instance in Derbyshire, been found with the skull, in such a manner as to leave no doubt of its having been placed in the mouth of the deceased. In some instances a considerable number of coins have been found deposited together, or scattered about, in a barrow, along with human remains. In Haddon Field a large number of coins, principally consisting of third brass of Constantine, Constans, Constantius II., Valentinian, Valens, and Gratian, were found, along with bones and fragments of pottery, traces of decayed wood, and a portion of a glass vessel. At Minning-Low, the fine chambered tumulus described on p. 54, *ante*, where several interments of the Romano-British period have undoubtedly been made in the earlier Celtic mound, many Roman coins, along with portions of sepulchral urns, etc., have from time to time been found. These are principally of Claudius Gothicus, Constantine the Great, Constantine Junior, Valentinian, and Constantius. In a barrow near Parwich, upwards of eighty coins of the later emperors were found. At most places, in fact, where Roman interments have taken place, coins have been found, and these range from an early to a late period in Roman history.

When interment was by inhumation, in many instances the body was simply laid in the earth without any further covering than the usual dress. In other instances there are abundant appearances of the body having been en-

* " Nec habet quem porrigat ore trientem."—JUVENAL.

closed in a wooden coffin or chest. In others, again, the body had been enclosed in a stone sarcophagus or chest, which was occasionally elaborately carved. Sometimes, again, coffins of lead were used. Mounds or barrows over sepulchral chambers and other modes of interment were frequently raised, to which I shall have to draw further attention. Examples of the first and most simple of these modes of burial have been discovered in different parts of the country, those at Bartlow Hills and at Little Chester being, perhaps, among the most notable. At the latter place a skeleton of a man found some years ago lay full length on its back, the arms straight down by the sides. Iron rivets, which were found much corroded, lay near various parts of the body, and a thin stratum of ferruginous matter encased the skeleton at a little distance from the body and limbs. From these circumstances it is to be inferred that the deceased was interred in his armour. Other interments by inhumation have been recently discovered in the same neighbourhood, but without, in some instances, the ferruginous appearances. The remains of horses were found along with them. Interments by inhumation have also been found at Brough and at other stations in the same county, and, as later deposits, in Celtic barrows. Those where the bones have been found *in situ* appear, like the one I have spoken of at Little Chester, at Bartlow, and at other places, to have been laid at full length on the back, the arms straight down by the sides. They appear in most instances to have been simply laid in a very shallow grave, but little below the surface of the already formed mound, and to have been then covered to no great thickness with earth. Those found at Bartlow lay parallel to, but a short distance apart from, each other, their heads to the west and feet to the east. They were laid flat on their backs, their limbs straight out, their arms by their sides, and hands on the thighs. Some coins of Constantine and Tetricus, and other little matters, were found with them.* Traces of wooden chests or coffins were discernible around these skeletons, and this feature is not uncommon in burials of this description.

When the body was placed in a stone chest or sarcophagus, it was in full dress, on its back, on the bottom of the chest, and any relics which were intended to be buried with it were laid about. The chest, as is evident from the examples found at York, was then partly filled with liquid lime, the face alone not being covered with the corroding liquid. When now found, a perfect impression of the figure is preserved in the bed of lime in which it was encased, and in some instances even the colour and texture of the dress is plainly distinguishable.† Frequently the stone chest contained a leaden coffin, in which the body was placed. A remarkably fine sculptured chest found in London,‡ and others found at York,§ will be sufficient references to these interesting sarcophagi, which are occasionally inscribed.

A tomb of a different description, which will be seen to partake largely of the construction of the stone cist of the earlier period, is here engraved (fig. 206). It is formed of ten rough slabs of gritstone, two on each side, one at each end, and four others laid as covering on the top. On removing the covering stones, a regularly shaped mass of lime presented itself, which had derived its form from a wooden coffin that had so nearly perished as to leave only small fragments behind. The wood was evidently cedar. On turning over this mass of lime an impression of the

* The skull of one of these, an excellent typical example of a Roman in the very prime of life, is engraved in "Crania Britannica," pl. 30.
† See example in the York museum.
‡ "Collectanea Antiqua," vol. iii., p. 45.
§ "Proceedings of the Yorkshire Philosophical Society;" Wellbeloved's "Eburacum;" "Crania Britannica," etc.

body of a man, which had been enveloped in, or covered with, a coarse linen cloth, fragments of which still remained, was distinctly seen. In another instance the impression of

Fig. 206.

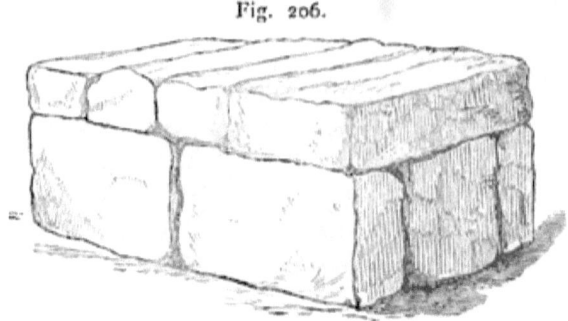

the body of a woman who had been clothed in rich purple, with a small child laid upon her lap, was distinctly visible in the lime.

Coffins of lead are of not unfrequent occurrence in the cemeteries of London, Colchester, York, Kingsholme,

Fig. 207.

Southfleet, Ozengal, and elsewhere. They are, as will be seen by the example from Colchester (fig. 207),* usually ornamented with raised escallop shells, beaded mouldings, annulets, etc., in a variety of ways. The next engraving (fig. 208) exhibits a leaden coffin discovered in 1864 at Bishopstoke,† in Hampshire. The lead which formed the

* Now in the Bateman Museum.
† See the *Reliquary*, vol. iv., p. 183.

coffin was about a quarter of an inch in thickness. The coffin, which was five feet six inches in length, and sixteen and a half inches in breadth, inside measure, had not been cast in a mould, but the lead cut so as to form the sides.

Fig. 208.

The lid appeared to be formed of one sheet, and had been bent or lapped over the lower part of the coffin. The lead was much corroded, and lime had evidently been placed in the coffin. There was none of the ornamentation on the outside, so common on leaden cists. Nearly the whole of the skeleton remained, but the skull was broken. The teeth were perfect and good. The skeleton was that of a female. Inside the cist were the remains of small glass bottles, probably lachrymatories. The glass was thin, and of a very pale green colour. There was no appearance of handles to the glass vessels, nor were there any marks of ornamentation on them, except a faint line or ring marked upon one of the three necks found. Around the coffin were the remains of the wooden chest in which it had been placed.

Coffins of baked clay, and cists formed of tiles, were also used. Of these, many examples have been found at York

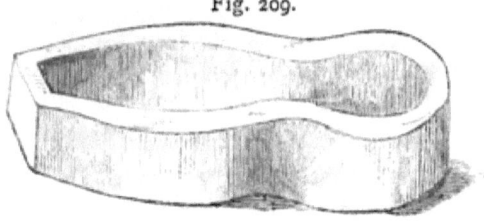

Fig. 209.

and at Aldborough. One of peculiar form, from the latter place, is here given (fig. 209).

Sepulchral chambers, sometimes of considerable size, were occasionally built above ground, and these were sometimes, like the immense chambered burial-places of earlier times, covered with a gigantic mound or barrow. A remarkable example of this is Eastlow Hill, in Suffolk, where the tomb appeared like a miniature house, of strong masonry, with the roof tiled and peaked. It was built upon a mass of concrete, the size of the tomb being twelve, by six and a half, feet. The walls were two feet thick, and the extreme height of the tomb, or house, was five feet. The interior was a cylindrical vault, and in the middle stood the leaden coffin containing the skeleton. The wooden chest in which it had been encased had decayed away, but some fragments and a number of nails remained. Over this remarkable tomb the mound called Eastlow Hill had been raised.

When the burial was by incremation, the ashes were carefully placed in the cinerary urn, and interred either by themselves, or with more or less ostentatious surroundings. In many instances a hole was dug in the earth, or in a Celtic barrow, and the urn, on being placed in it, simply covered with a flat stone. At other times it was placed in a sarcophagus, or chest, and surrounded with vessels of various kinds and with other relics. At others, again, it was enclosed in a leaden, or stone, or other vessel, before being consigned to the earth. In many cases barrows were raised over these remains. There was a general belief in the minds of the Roman people that articles of various kinds buried or burnt with their dead, would add to the comfort and happiness of the spirit in another world. Thus jewels, personal ornaments, food, wine, articles for the toilet, culinary vessels, pottery, and glass of various kinds, and numberless other articles were buried or burned with the bodies. Branches of trees and garlands were also burned or buried with the dead.

Some remarkable examples of tombs and graves containing burials by cremation have been discovered at the Bartlow Hills, at Colchester, at Uriconium, at Rochester, at York, at Chester, and in other places. The grave, or chest, was formed of wood, or tiles, or of stone. In this the urn containing the ashes of the dead was placed, and around it were put smaller vessels which probably contained ointments, balsams, and other offerings; a lamp; and other articles. One example, formed by tiles, contained when discovered a few years ago, besides fragments of the cinerary urn, four earthenware bottles, six paterae, three small urn-shaped vessels, a terra-cotta lamp, and a lachrymatory. A chest of stone (fig. 210) found at Avisford,

Fig. 210.

Sussex, contained a large square vase of fine green glass, filled with burnt bones, and around it were placed three elegant vases with handles, several paterae, a pair of sandals elegantly and fancifully studded with brass nails, an oval dish with handle containing a fine agate, a double-handled glass bottle, and three lamps placed on projections in the angles of the chest.

An example of a tomb formed of tiles is shown on the next engraving (fig. 211). It was found at York, and is composed of ten roof tiles, with a row of ridge tiles at the top. Within this the interment had taken place. The tiles were inscribed with the impress of the Sixth Legion—

Fig. 211.

LEG VI VI (*Legio Sexta Victrix*—"the Sixth Legion victorious.")

Sepulchral inscriptions to the memory of the deceased are not uncommon, and one or two examples of their style of wording will be sufficient. One, at York, reads thus:—

   D . M . SIMPLICIAE . FLORENTINE
   ANIME . INNOCENTISSIME
   QVE . VIXIT . MENSES . DECEM
   FELICIVS . SIMPLEX . PATER . FECIT
   LEG . VI . V.

"To the gods of the shades. To Simplicia Florentina, a most innocent thing, who lived ten months. Her father of the Sixth Legion, the victorious, made this." Another, from Carvoran in Northumberland, is thus affectionately worded:—

    D . M
    AVRE . FAIAE
    D . SALONAS
    AVR . MARCVS
    C . OBESEQ . CON
    IVG . SANCTIS
    SIMAE . QVAE . VI
    XIT ANNIS XXXIII
    SINE VLLA MACVLA

"To the gods of the shades. To Aurelia Faia, a native of Salona, Aurelius Marcus, a centurion, out of affection for his most holy wife, who lived thirty-three years without any stain." Another, from Caerleon, is thus :—

    D . M . IVL . IVLIANVS
    MIL . LEG . II . AVG . STIP
    XVIII . ANNOR . XL
    HIC . SITVS . EST
    CVRA . AGENTE
    AMANDA
    CONIVGE

"To the gods of the shades. Julius Julianus, a soldier of the Second Legion, the Augustan, served eighteen years, aged forty, is laid here by the care of Amanda his wife." Another, from Chesters, in Northumberland, is as follows:—

    D . M . S
    FABIE HONOR
    ATE . FABIVS . HON
    ORATIVS . TRIBVN
    COH . I . VANGION
    ET . AVRELIA . EGLIC
    IANE . FECER
    VNT . FILIE . D
    VLCISSIMME

"Sacred to the gods of the shades. To Fabia Honorata, Fabius Honoratius, Tribune, of the First Cohort of Vangiones, and Aurelia Egleciane, made this to their daughter most sweet." And one at Bath is thus :—

    D . M.
    AEL . MERCV
    RIALI . CORNICVL
    VACIA . SOROR
    FECIT

"To the gods of the shades. To Ælius Mercurialis, a trumpeter, his sister Vacia made this."

The articles which the grave-mounds and cemeteries of the Romano-British period most frequently produce are pottery of various kinds; glass vessels; coins; arms, both of bronze and of iron; fibulæ, armillæ, and other personal ornaments; knives, scissors, etc.; and a large variety of other things. To a brief notice of these contents of the graves I shall next, in this division of my work, confine myself.

CHAPTER VIII.

Romano-British Period—Pottery—Durobrivian Ware—Upchurch Ware —Salopian Ware—Pottery found at Uriconium—Potteries of the New Forest, of Yorkshire, and of other places—Sepulchral Urns— Domestic and other vessels.

THE pottery of the Romano-British period, so far as relates to what is found in the grave-mounds of that people, consists, in the main, of cinerary urns, jugs (so called), pateræ, amphoræ, bowls, and vases of various kinds. Of the pottery alone of this period, sufficient interesting matter to fill a couple of goodly volumes might easily be written. It will, therefore, be at once understood that in a work like the present, which is simply intended to be a descriptive sketch of the contents of grave-mounds, elaborate accounts of the different kinds of ware made by that people, and of the modes of manufacture which they adopted, would be unnecessary. The principal divisions are the Samian ware, the Durobrivian ware, the pottery of the Upchurch marshes, the Hampshire ware, the Salopian ware, and the Yorkshire wares, and to these divisions I shall devote some few pages, and in so doing express thanks to my friend, Mr. Thomas Wright, for some excellent articles* on the Durobrivian, the Upchurch, and the Samian wares, which he has written. Before proceeding to speak of the different vessels found with interments, it will be well to glance at these different wares and their characteristics.

* In the *Intellectual Observer*.

The *Durobrivian* or *Castor ware*, as it is variously called, is the production of the extensive Romano-British potteries on the river Nen in Northamptonshire and Huntingdonshire, which, with settlements, are computed to have covered a district of some twenty square miles in extent. The discovery of this pottery and of the kilns in which its productions were fired, etc.—one of which is engraved on fig. 212—is due to the late Mr. Artis,

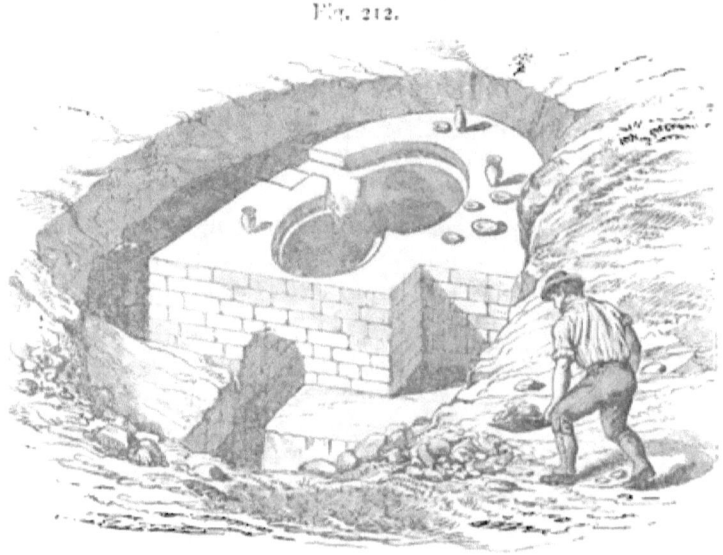

Fig. 212.

who prosecuted his examination of the locality with great perseverance and skill. There are several varieties of this Durobrivian ware, and two especially have been remarked; the first, blue or slate-coloured, the other reddish-brown or of a dark copper colour. The former was coloured by a simple though curious process, which Mr. Artis was enabled to investigate in a very satisfactory manner. It will, perhaps, be best told in his own words. " During an examination of the pigments used by the

Roman potters of this place," he says, " I was led to the conclusion that the blue and slate-coloured vessels met with here in such abundance, were coloured by suffocating the fire of the kiln at the time when its contents had acquired a degree of heat sufficient to ensure uniformity of colour. I had so firmly made up my mind upon the process of manufacturing and firing this peculiar kind of earthenware, that for some time previous to the recent discovery [in 1844] I had denominated the kilns in which it had been fired *smother kilns*. The mode of manufacturing the bricks of which these kilns are made is worthy of notice. The clay was previously mixed with about one-third of rye in the chaff, which, being consumed by the fire, left cavities in the room of the grains. This might have been intended to modify expansion and contraction, as well as to assist the gradual distribution of the colouring vapour. The mouth of the furnace and top of the kiln were, no doubt, stopped; thus we find every part of the kiln, from the inside wall to the earth on the outside, and every part of the clay wrappers of the dome, penetrated with the colouring exhalation. As further proof that the colouring of the ware was imparted by firing, I collected the clays of the neighbourhood, including specimens from the immediate vicinity of the smother kilns. In colour some of these clays resembled the ware after firing, and some were darker. I submitted them to a process similar to that I have described. The clays dug near the kilns whitened in firing, probably from being bituminous. I also put some fragments of the blue pottery into the kiln; they came out precisely of the same colour as the clay fired with them, which had been taken from the side of the kilns. The experiment proved to me that the colour could not be attributed to any metallic oxide, either existing in the clay or applied externally; and this conclusion is confirmed by the appearance of the clay wrappers of the dome of the kiln.

It should be remarked, that this colour is so volatile that it is expelled by a second firing in an open kiln." Fortunately, some of the kilns remained almost entire, and many had been left with the pottery partly packed in them for firing, so that there was no difficulty in understanding the nature of the process here employed by the Roman potters.

This Durobrivian pottery is especially interesting, from its being covered with ornaments and figures, in relief,

Fig. 213.   Fig. 214.   Fig. 215.

like those on the Samian ware, but not like it cast from moulds. "The vessel," Mr. Artis remarks, "after being thrown upon the wheel, would be allowed to become somewhat firm, but only sufficiently so for the purpose of the lathe. In the indented ware, the indenting would have to be performed with the vessel in as pliable a state as it could be taken from the lathe." The ornamenter then took a slip of rather liquid material, and with an implement made for the purpose, formed all the ornaments and figures with the hand. The slip used for this purpose was often white, which was laid on a dark ground. "The vessels, on which are displayed a variety of hunting subjects, representations of fishes, scrolls, and human figures, were all glazed after the figures were laid on; where, however, the decorations are white, the vessels were glazed before the ornaments were added. Ornamenting with figures of animals was effected by means of sharp and blunt skewery instruments, and a slip of suitable consistency. These instruments seem to have been of two kinds: one thick enough to carry sufficient slip for the nose, neck, body, and front thigh; the other of a more delicate kind, for a thinner slip for the tongue, lower jaws, eye, fore and hind legs, and tail. There seems to have been no retouching after the slip trailed from the instrument."

Of the forms of mere ornamentation of this ware, the scroll ornaments appear to have been the most popular. The arrangement and combination of the scrolls, which are sufficiently varied, are often both tasteful and very effective.

Fig. 216.

In the cut (fig. 216) I have selected two examples of the most common forms of this kind of ornamentation, and others I show on the following engravings, figs. 217, 218, and 219, and again on figs. 213, 214, and 215.

" It is, however, the figured pottery of Durobrivæ, which

presents some of the characteristics of the Samian ware, that possesses the greatest interest for the antiquary and the historian. The variety of subjects in the Samian ware is far greater, and they are treated in a more elaborate and more highly finished style of art, yet similar classes of subjects appear to have enjoyed greater popularity than others in the Durobrivian and Samian pottery, and we can hardly

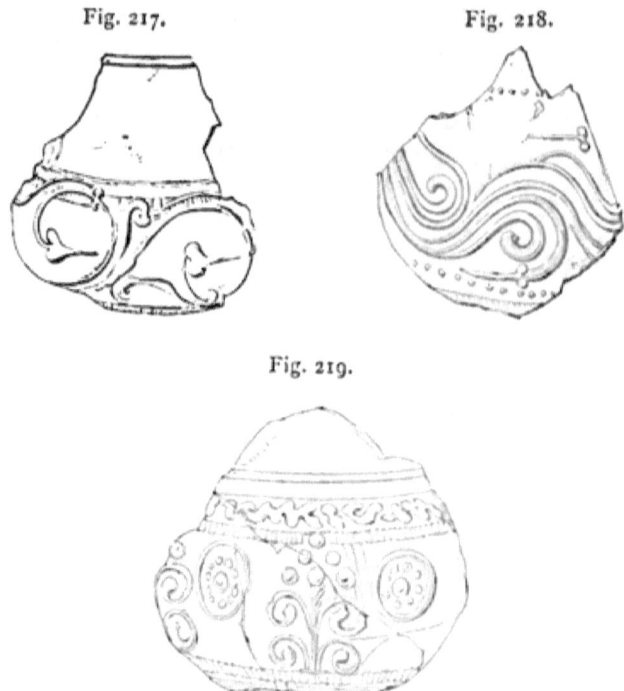

Fig. 217.   Fig. 218.

Fig. 219.

help suspecting that there was some design of imitating, or perhaps a sentiment of rivalry. Considering that they were only executed with the hand, and it would appear rapidly, the style of drawing is remarkably good and spirited. But they have another and a peculiar value;

when we consider that they were certainly executed in this country, and by artists who could hardly have done otherwise than copy what was constantly before their eyes, we can have no doubt that these are all true pictures, pictures which we could hardly in any other way have obtained, of life in Britain under the Romans, and they show us, as well as could be shown in subjects capable of being represented by such artists, those occupations in which the enjoyment of life was then believed to consist. The more common of these subjects are hunting scenes and scenes taken from the amphitheatre or racecourse." For instance, the dog hunting the hare, given in our cut (fig. 220) taken from an example of Durobrivian ware engraved

Fig. 220.

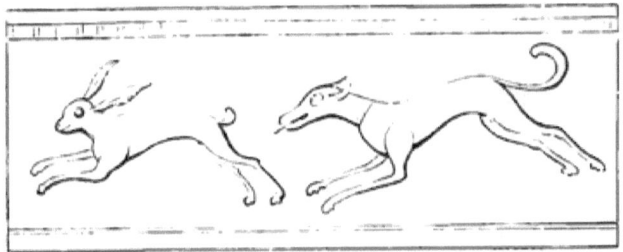

in Artis's plates, must be recognized at once as a greyhound, the same variety of dog which is still used for the same purpose. It has been suggested that this may be the dog to which the Romans gave the name of *vertagus*, and which is said to have been a British dog. Martial describes it in one of his epigrams as—

" Divisa Britannia mittit
Veloces, nostrique orbis venatibus aptos."
Nemesiani *Cynegetica*, l. 123.

Other examples of hunting subjects are here given (figs.

221 to 225), and others again will be found on a subsequent page.

Fig. 221.   Fig. 222.

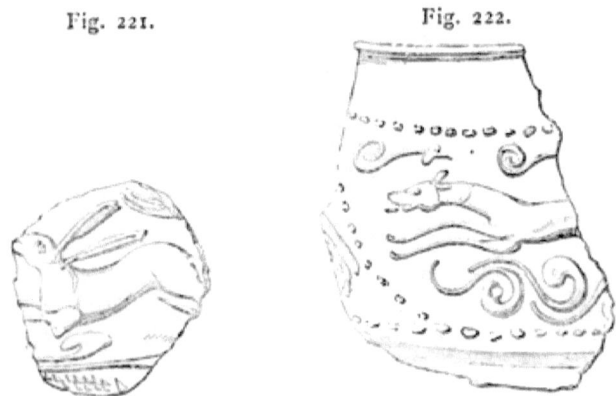

The engravings fig. 223 and 224, taken from a sample of

Fig. 223.

Fig. 224.

this pottery given in one of Mr. Artis's plates, represents the British staghound of the Roman period chasing a stag. We have a different dog in other examples, as in fig. 225, which is taken from a very remarkable vessel of this ware, now known as the Colchester vase, where it appears driving before it both stags and hares. The hunting of the boar is

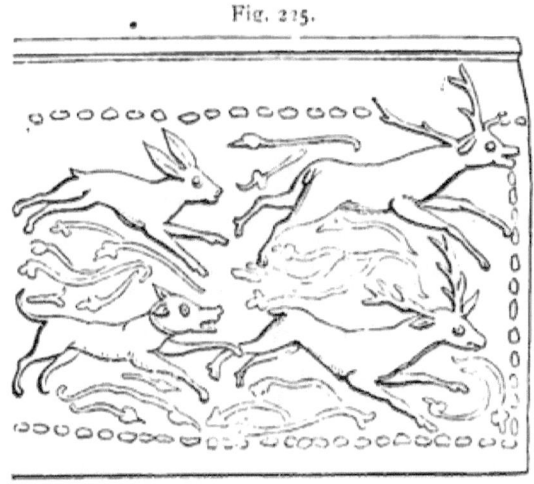

Fig. 225.

also introduced in some examples of this pottery. Gladiatorial combats are also favourite subjects on the pottery made at Durobrivæ, as on the Samian ware, and they leave no doubt that these cruel and degrading exhibitions were cherished by the Romans in Britain as well as in Italy.

That very remarkable monument of the ceramic art in Roman Britain, known as the "Colchester vase," was found in 1853, in the Roman cemetery which occupied the site of West Lodge, near Colchester. It had been used as a sepulchral urn, and when found contained calcined bones, and was covered with an inverted shallow vessel or dish. "The ornamentations consist of three groups, one of which is the flight of stags and hares pursued by a dog, given in our cut (fig. 225). The second and, perhaps we may say, the principal group represents, in perfectly correct drawings, the combat of the two classes of gladiators, a *Secutor* and a *Retiarius*, the latter of whom, vanquished, has dropped his trident, and raises his hand to implore the mercy of the spectators. The *Secutor*, with a close helmet over his head, and a short sword in his hand, advances to strike the fatal blow, unless arrested by the success of his adversary's appeal. Over the head of the *Retiarius* is the inscription, VALENTINV LEGIONIS XXX., meaning clearly, "Valentinus, of the thirtieth legion," which was doubtless the name of the individual here represented. A similar inscription over the head of the *Secutor* is read without difficulty —MEMN.N.SAC.VIIII., which is explained by Mr. Roach Smith, who considers the A in SAC as an error for E, as standing for *Memnius* (or *Memnon*) *numeri secutorum victor ter; i.e.*, "Memnius, or Memnon, of the number (or band) of secutors, conqueror thrice." There is no reason for supposing that this inscription has any reference to the individual whose remains were buried in the vase, but it has probably reference to some remarkable gladiatorial combat which had created a sensation in Roman Britain, like some one of the celebrated boxing matches of modern times; sufficiently so to have become a popular subject of pictorial representation.*

"The third group on the Colchester vase also represents a performance which was very popular among the Romans and among Saxons, and, indeed, throughout the Middle Ages, that of a bear-tamer and disciplined bear. The bear, in this case, appears inclined to be rebellious, and his keeper, whose left arm bears what appears to be a shield, and his legs and right arm protected by bands or thongs, is menacing the animal with a whip. An assistant is approaching, with what appear to be two staves in his hands, for the purpose also of intimidating the ferocious animal. Over the

* See Mr. Roach Smith's interesting account of this vase in the "Collectanea Antiqua," vol. iv., pp. 82-89.

head of the man holding the whip are the letters SECVNDVS MARIO, the intended application of which is not very clear."

On another vase in the British Museum, the figures represent a chariot-race in the Roman racecourse or stadium.

Another class of subjects of extreme interest, as coming from a Romano-British pottery, are mythological subjects, which appear to have been rather a favourite ornament of the Durobrivian pottery. Fragments of several vessels, with the figures of the seven gods and goddesses, have been met with. Another characteristic of the Durobrivian

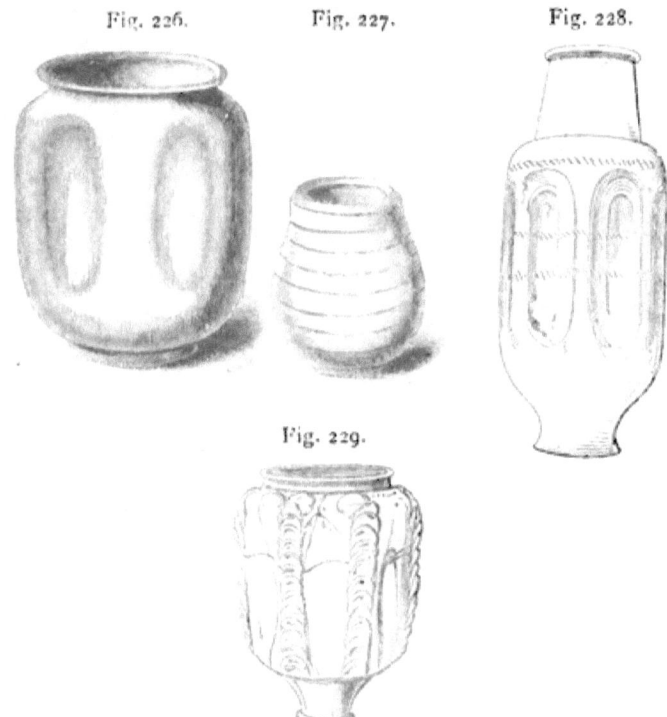

Fig. 226.   Fig. 227.   Fig. 228.

Fig. 229.

ware, consists of indentations made in the side of the

* Wright.

vessel, while still soft, but after it had left the lathe, and continued with regularity round it. Sometimes, where little ornament was employed on the rest of the vase, these indentations were left quite plain; sometimes an ornament was introduced in the centre; and not unfrequently the indentation was formed into a niche for the reception of a figure. For indented vases see figs. 226, 228, and 229.

The *Upchurch ware*, so called because made on the tract of land now known as the Upchurch Marshes, on the river Medway below Chatham, is next in importance, as far as extent of works go, to the Castor ware. The district where these pot-works are proved to have existed extends to a distance of five or six miles in length, and from one to two miles in breadth, and throughout this tract a bed of refuse pottery exists. This is seen to the best advantage about Otterham Creek, not far from Upchurch church, and from its being first noticed here the name of Upchurch ware has arisen.

"The Roman ware made in the Upchurch potteries presents distinctive peculiarities which cannot be mistaken, and it must have been in great repute, certainly the next after the foreign Samian and the native Durobrivian wares, in this province of the empire. Like the Durobrivian, too, it has been found on Roman sites in France and Germany, so that it was probably exported. As Battely has described it, the greater proportion of this ware is of a 'blackish colour,' or rather of a bluish or greyish black, which was produced, no doubt, by the process of the smother-kiln, already described in connection with the Durobrivian pottery. Some of the Upchurch pottery presents a colour approaching to dark drab. Examples of both are given. The forms, as well as the sizes, vary greatly, but they all present those delicate forms of the curve which we recognise at once as coming from the hands of the Roman artist. The texture of the pottery itself is fine, and it is very thin. The ornamentation also is varied, but not very elaborate or very refined. One of the patterns consists of a band of half-circles, made with compasses, from each of which a band of parallel lines descends vertically. Examples of various kinds of ornament are given in the accompanying woodcut

Fig. 230.

(fig. 230). The little vessel in the front of the cut has had two handles, but one is lost; it is supposed to be an incense pot.

"The instruments used in the ornamentation of this pottery appear to have been of a very rude description, and were, as it seems, chiefly mere sticks, some sharpened to a point, and others with a transverse section cut into notches. The former were used in tracing the lines already described; the latter had the section formed into a square, or rhomboid, the surface of which was cut into parallel lines crossing each other, so as to form a dotted figure, and this was stamped on the surface of the pottery in various combinations and arrangements. Sometimes these dots are arranged so as to form bands, as in the example in the back of the group. The large urn in the middle of the group furnishes an example of another kind of ornamentation found on the Upchurch pottery, formed by parallel inter-

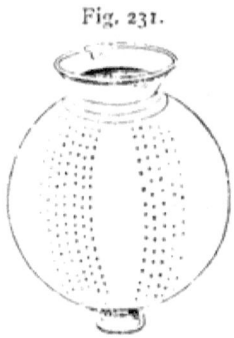

Fig. 231.

secting lines. In its shape this vessel has much the appearance of a sepulchral urn. A considerable quantity of this pottery is without ornament at all. Among this unornamented pottery are found, especially, jug-shaped vessels, commonly with a handle. Two of these vessels are represented in the group, in which is also shown a curiously shaped plain urn and an unornamented vessel of another form. At different spots over the locality which was covered by these potteries, Mr. Roach Smith has found remains which indicate the former existence of kilns, and further researches will most probably bring to light some of the kilns themselves. Traces have also been found of the residences and of the graves of the potters."*

The *Romano-Salopian* potteries—the works which produced such a large quantity of vessels from the clay of the Severn valley, probably in the neighbourhood of Broseley, which bed is still worked for fictile purposes—were, there

* Thomas Wright.

is reason to believe, much less extensive than either of those spoken of, but yet they must, from the large quantity of examples which have been dug up at Wroxeter, have been of some considerable extent. Of these wares, "two sorts especially are found in considerable abundance; the one white, the other of a rather light red colour. The white, which is made of what is commonly called Broseley clay, and is rather coarse in texture, consists chiefly of rather handsomely shaped jugs of different sizes, of mortaria, and of bowls of different shapes and sizes, which are often painted with stripes of red and yellow. The other variety, the red Romano-Salopian ware, is also made from one of the clays of the Severn valley, but it is of a finer texture, and consists principally of jugs not dissimilar to those in the white ware, except in a very different form of mouth, and of bowl-shaped colanders."* A group of vessels of the Salopian ware is here given (fig. 232). These examples are all from Uriconium (Wroxeter), and have been found in the cemetery there. They are cinerary urns which have, of course, contained the ashes of the dead, and domestic vessels which have been buried along with them.

The pottery of the New Forest bears in some respects a striking resemblance to that from Castor. The clays there found were white and fawn-coloured.† The Yorkshire productions present some peculiarities in pattern which will be noticed later on, and those of Oxfordshire are somewhat similar to the Castor ware. Of other pot manufactories it will not be necessary to speak in this work.‡

The sepulchral urns—those which were intended to receive the burnt bones of the dead—vary much in size as well as in form, material, and ornamentation. Many are of

---

* Wright.
† For an interesting account of these potteries, see Wise's "New Forest."
‡ For a detailed account of all the different pot-works and their productions, see my "Ceramic Art in England."

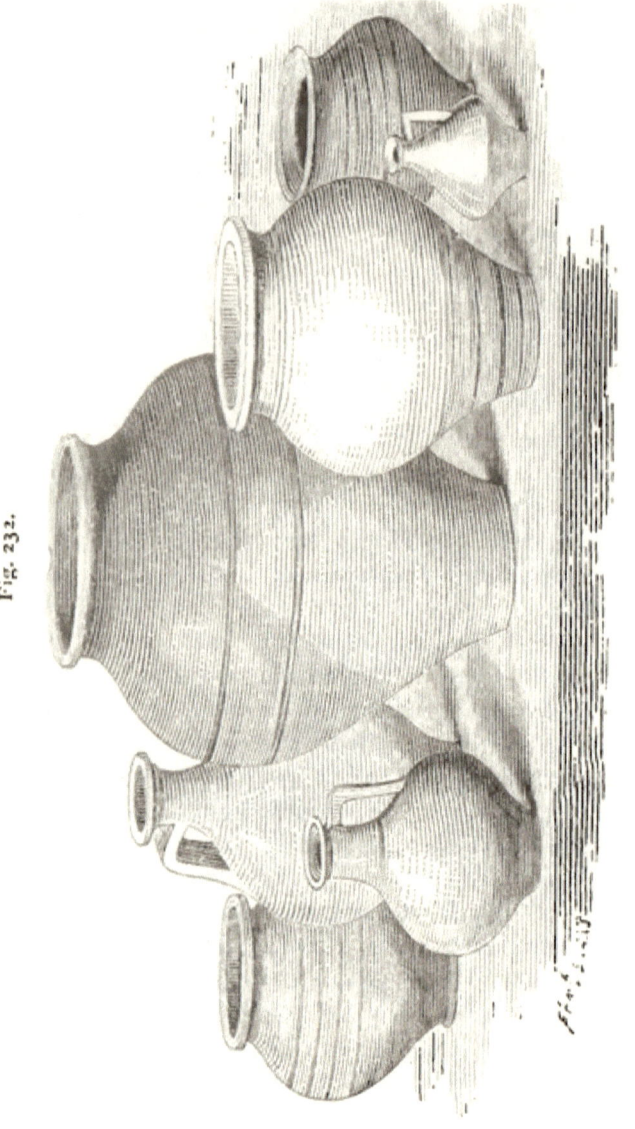

Fig. 232.

globular form, and of a dark bluish-grey colour in fracture They are somewhat coarse in texture, and are thrown on

Fig. 233.

the wheel. The engraving (fig. 234) exhibits one of these vessels. When found, it was, like the others I am about

Fig. 234.

to notice, filled with burnt bones. The engravings (figs. 235 and 236) show two urns containing human remains, the smaller one of which, found at Little Chester, is formed

Fig. 235.

of a black clay, mixed with small pieces of broken shells—a kind of pottery much used for sepulchral purposes. The

Fig. 236.

larger urn is of a hard and compact clay, and is beautifully "thrown" on the wheel. These examples are entirely devoid of ornament. A good example of this form will be seen in the centre of the group (fig. 230), but in this instance the urn is covered with a reticulated ornament. Examples whose forms partake a little more of the jar shape will be noticed on fig. 232, and others are given on Fig. 233, Nos. 1, 2, 3, and 6. Fig. 237 is from Little Chester, and is

Fig. 237.

formed of a fine reddish-brown clay, and is ornamented with "slip" in an unusual manner. It measures 3½ inches only in height, and the same in diameter at the mouth.

Fig. 238.   Fig. 239.

 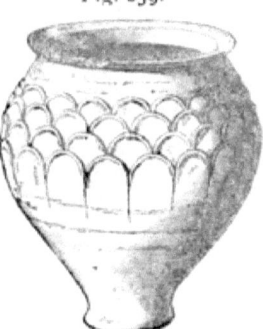

When found, it was filled with burnt bones, among which were some small fragments of bronze ornaments, which had evidently been burned along with the body. The next

Fig. 240.   Fig. 241.   Fig. 242.

Fig. 243.   Fig. 244.

Fig. 245.   Fig. 246.   Fig. 247.

SEPULCHRAL URNS.

examples (figs. 238, 242, 248, and 249), are of a different character, both in ornamentation and in colour of clay.

The domestic vessels, and other varieties of Roman pottery found with interments, vary very considerably one from another, so much so, indeed, as almost to require

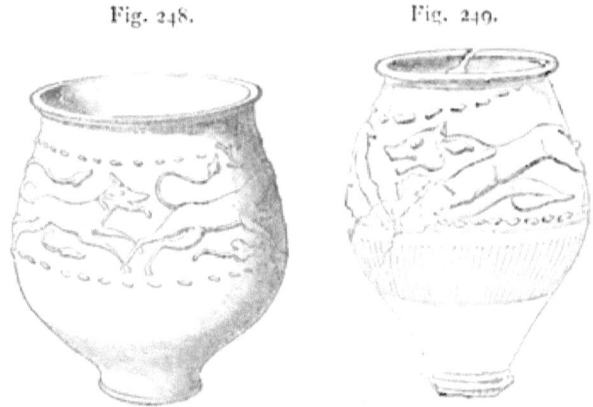

Fig. 248.   Fig. 249.

a detailed dissertation on the entire fictile arts of that people. Examples of some of the different vessels which are found are shown on figs. 243 to 266, and on figs. 230, 232, and 233, which exhibit some of the more usual and better

Fig. 254.

known forms. Figs. 250, 252, and 253 are amphoræ, found in London, as was also the small amphora-shaped vessel, fig. 251. Fig. 254 is a good typical example of a morta-

Fig. 250. Fig. 251.

Fig. 252. Fig. 253.

rium, of which considerable numbers, usually in fragments, are found wherever there has been a Roman settlement. The next group (fig. 255) represents five examples of black-ware vessels, the ornaments on which are produced by

Fig. 255.

tracing lines on the surface. The remainder of the engravings (figs. 258 to 268) exhibit cups, bowls, unguentaria

Fig. 256.   Fig. 257.   Fig. 258.

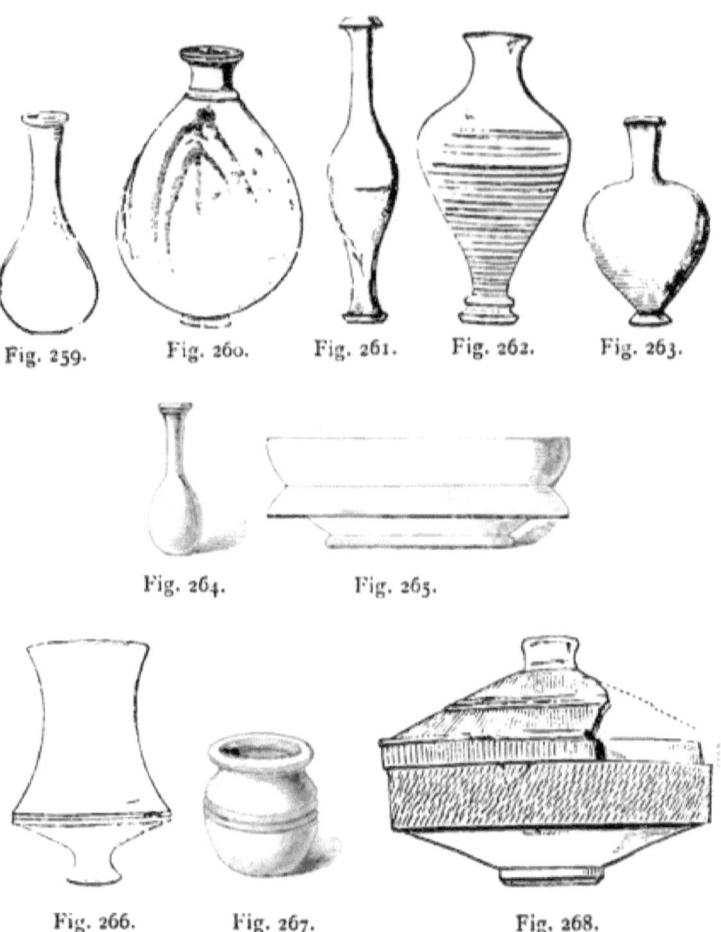

Fig. 259. Fig. 260. Fig. 261. Fig. 262. Fig. 263.

Fig. 264. Fig. 265.

Fig. 266. Fig. 267. Fig. 268.

of different forms, and various shapes of vases. They are all characteristic examples of Romano-British ware, and will be useful to the student in correctly appropriating any specimens which may fall into his hands.

CHAPTER IX.

Romano-British Period—Pottery—Samian Ware — Potters' Stamps—
Varieties of Ornamentation—Glass Vessels—Sepulchral Vases, etc.—
Lachrymatories—Bowls—Beads—Coins found with Interments.

IN the preceding chapters I have purposely avoided including vessels of Samian ware. As these are frequently found with sepulchral deposits, I now proceed to speak of this peculiar and beautiful ware.

Samian ware is that peculiarly fine, close-textured, and richly-coloured red-ware, which is so frequently found, and is so well known to antiquaries. The body of this ware is of a fine red colour, but its surface is of a deeper and richer tone, much like the best red sealing-wax. It is extremely

Fig. 269.

hard and brittle, and is sonorous in sound when struck. The vessels of this ware consist for the most part of bowls, cups, and pateræ or dishes, in each of which divisions are found an almost endless variety of forms, and while some are perfectly plain, others are more or less covered with ornaments—figures of men, animals, foliage, borders, etc.,—in relief. These relief ornaments were produced from moulds, and the names of the makers of the vessels were also frequently stamped upon them. Of these ornaments and potters' marks, Mr. Wright says, "The potter's name was placed in a small rectangular label, as in the examples given

Figs. 270 and 271.

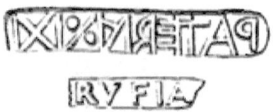

on figs. 270, 271, 272, 273, and 274. The name was most com-

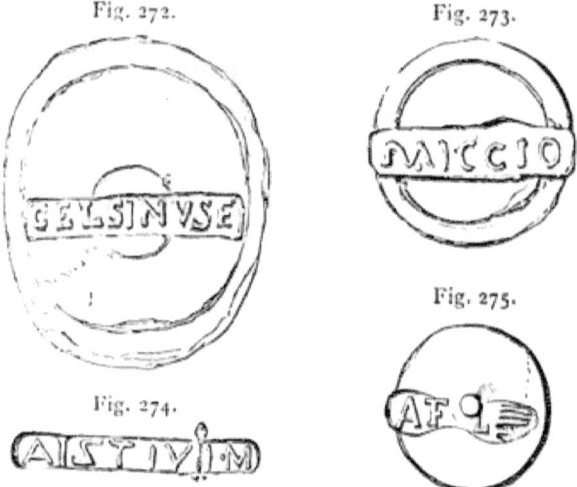

Fig. 272.

Fig. 273.

Fig. 275.

Fig. 274.

monly put in the genitive case, combined with O or OF, abbreviations of the word *officina*, as in the example given in our cut, where OF MODESTI stands for *officina Modesti*, i.e. 'from the workshop of Modestus;' or with M for *manu*, as COBNERTI M, for *Cobnerti manu*, 'by' or 'from the hand of Cobnertus." Sometimes the name is given in the nominative case, followed by F or FE, for *fecit*, as COCVRO F, for *Cocuro fecit*, 'Cocuro made it.' Doubled or ligulated letters are frequently introduced into these inscriptions, an example of which is given in the lower figure to the right, where the first letter is the ligulated T and E, and the name is TETTVR. Sometimes we meet with an error in the spelling of the word; and in one or two instances the person who made the stamp inscribed the name carelessly, so that it read direct on the stamp, and consequently it is reversed in the impression on the pottery. An example is given in the cut, where the inscription reversed reads PRASSO·O. The name is not always placed in a square label, though examples to the contrary are rare. In a few instances it has been found inscribed round a small circle. It is a peculiarity of the Arrentine ware, described by Fabroni, that the label not unfrequently assumes the form of the sole of a man's foot. The stamp of this form given in our cut occurs on a piece of the red Samian ware found at Lillebonne, in Normandy. The inscription appears to be HIL · O · L · TITI, which may perhaps stand for *Hilarii officina liberti Titi*, 'from the workshop of Hilarius, the freedman of Titus.' The next cut (fig. 276) represents one of the stamps used for impressing the label with the potter's name. It was found at Lezoux, in Auvergne and presents the name AVSTRI·OF.

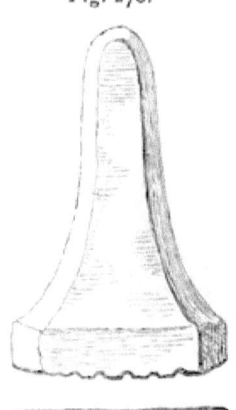

Fig. 276.

'from the workshop of Auster.' A similar die of a potter named Cobnertus is preserved in the museum at Sevres. Both these names occur on specimens of Samian ware found in England. Other potters' names are shown on figs. 272 to 275. The first of these bears the name CELSINVS . F.; the second, MICCIO; the third, AISTIVI . M ; and the fourth is the one referring to Aretium."

Similar dies for stamping the ornaments and figures have also been found in France. In the latter, each die contained a single figure, or, at all events, a single group, and this explains why the same figures are so frequently found on the pottery in different combinations. One of these dies contains a single festoon and tassel of the well-known festoon ornament, so common on this pottery.

The ornamental borders which are most commonly met with on Samian ware are elegant festoon-and-tassel borders, and the egg-and-tongue ornament, both of which, as well as a border consisting of a range of figures representing the Medicean Venus, are shown on the accom-

Fig. 277.

panying engraving (fig. 277) of a fine bowl found in London. Wavy lines and lines of circles are also common, and we frequently meet with scroll-work of very elegant design, commonly formed of leaves, flowers, and fruit.

SAMIAN WARE—BOWLS. 179

Examples, selected from a numerous variety, are given on

Fig. 278.

Fig. 279.

engravings (figs. 278 to 282.) These scrolls are generally

Fig. 280.

used to form a border round the upper part of the bowl, as shown on figs. 280 and 281. The foliage most in favour

Fig. 281.

for these scrolls was that of the vine, and the ivy (fig. 282),

Fig. 282.

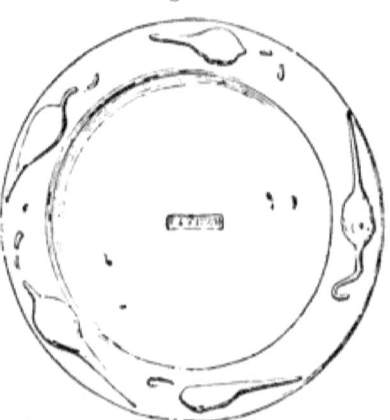

and also that of the strawberry; the former of which especially shows that this pottery was, as Pliny says of the Samian ware, particularly intended for the service of the table. The ivy-leaf, indeed, is almost the only ornament of the plainer description of this red ware. Sometimes the leaves of the vine are gracefully intermingled with the clusters of the fruit, and with little birds which are feeding upon the latter, as in the fragment represented in our cut (fig. 283).

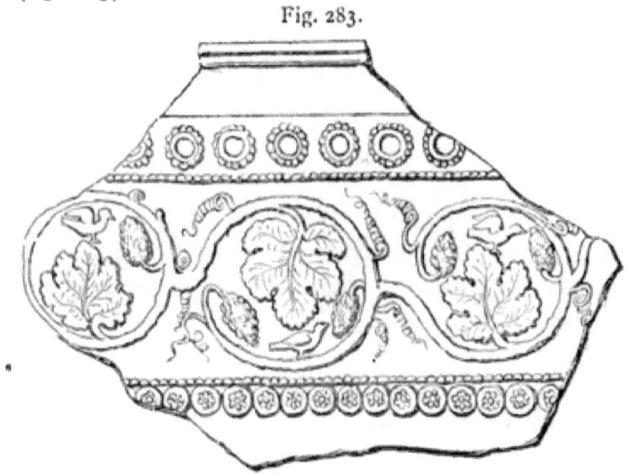

Fig. 283.

Animals of all kinds are found in abundance among the ornaments of the Samian ware. Among these the boar was a great favourite. For instance, a cup will be divided into compartments, in which figure alternately two boars, and a man confronting them with a spear. In a similar compartment under arches, in another, we have two heads of lions above, and below, a rabbit and a dog. Another, again, is ornamented with fishes, separated by squares filled with a singular ornament, which is perhaps intended to represent water. Sometimes the whole outside of a bowl is covered with birds, beasts, and fishes mixed together in the utmost confusion.

The subjects in which human figures are introduced present still greater variety, and it need hardly be added that they are much more interesting. Subjects from the classical mythology are very common, and among the figures of the deities we recognize some, such as the Venus de Medici (fig. 277), which were copied from well-known models of art. Combats of pygmies and cranes appear as favourite subjects, as in the

Fig. 284.

paintings, etc., in Pompeii. Sacrifices and religious ceremonies are not uncommon; and especially bacchanalian processions, and dances of bacchantes and satyrs—another proof that this ware was used for the festive board. The spirited manner in which figure subjects are often treated, will be seen by the engravings we have given, and by examples to be found in most collections. One vessel

Fig. 285.

represents a bacchanalian scene, in which Silenus figures among satyrs and fauns. A faun is drinking from a horn supplied from a wine skin which he holds in his left hand, while Silenus attempts to snatch it from his hands. Genii, one of whom appears with wings on another fragment of the same vessel, appear to be directing or presiding over the scene. Among other very favourite subjects are hunting scenes, gladiatorial combats, and the sports of the amphitheatre. Others represent sacrifices and religious offerings. Musicians performing on various instruments

Fig. 286.

are also common; and domestic scenes are depicted in great variety. Many of these are of a character not to

Fig. 287.

be described, but sufficiently characteristic of the degraded state of morality under the Roman empire. The bowls here engraved (figs. 285 and 286) are good examples of these kinds of decoration. Another is ornamented with a series of figures, which appear to have no connection one with another. In the middle is a bacchanal with his thirsus; to the right of him a figure playing on a double pipe; on both sides a group of bears; and to the extreme right a charioteer, followed by a bear "rampant."

The great quantity of this Samian ware which is found on Roman sites admits of easy explanation, from the circumstance that it was held in great favour, and that the manufactories on the Continent continued to work with activity in producing it during the whole Roman period. The number of names of potters, collected from fragments found in England alone, amounts to more than two thousand, and we must suppose them to have been spread over a long period.

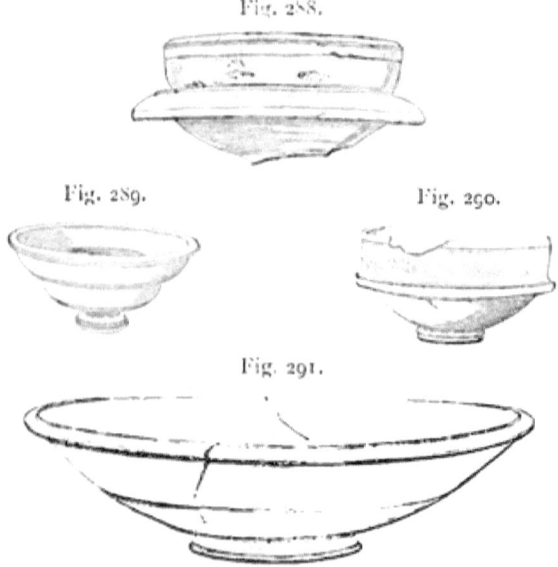

Fig. 288.

Fig. 289.   Fig. 290.

Fig. 291.

Other examples of the common forms of Samian-ware vessels are given on figs. 288 to 291, and a clay mould for forming heads on pottery, discovered by myself at Headington, is shown on fig. 287.

Glass was very successfully and beautifully worked by the Romans, not only abroad, but in Britain, and vessels of this material are frequently found with sepulchral deposits. They are of great variety, and evidently made for many different uses. Those found in the graves are usually those made for holding the burnt bones of the dead; small vessels, commonly called lachrymatories, although their use was most probably that of holding the unguents and aromatics usually buried with the dead; small bowls, cups, or drinking vessels; and beads.

Of the sepulchral vessels of glass the one here engraved (fig 292), from Bartlow Hills, will show the general form.

Fig. 292.

They are of somewhat thick green glass, with neck and handle, and are literally bottles. The one from Bartlow Hills is of square form, and is six inches in height and four inches square on the bottom. Others are round in form. They contained the calcined bones of the dead. Of the forms of the small vessels known as lachrymatories, to which I have alluded, the examples in pottery on figs. 259 to 263 will convey a tolerably correct idea. They are usually from three to five inches in height. One found at Mount Bures,

Colchester, is a remarkable example, being made of beautifully variegated glass. Cups or bowls, or, as they may not inaptly be called, basins, are of the common basin form, or jar shaped. They are usually of green glass, and of

Fig. 293.

elegant workmanship. Beads are, perhaps, the most frequently found of any remains of Roman glass; this being of course owing to their more solid and, consequently, less

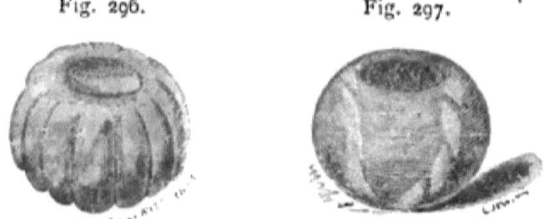

Fig. 294.   Fig. 295.

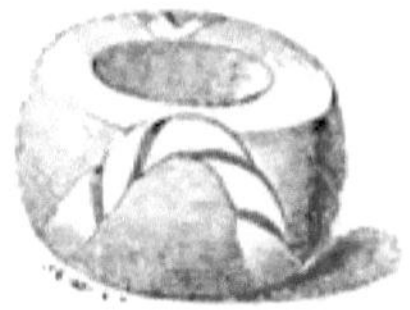

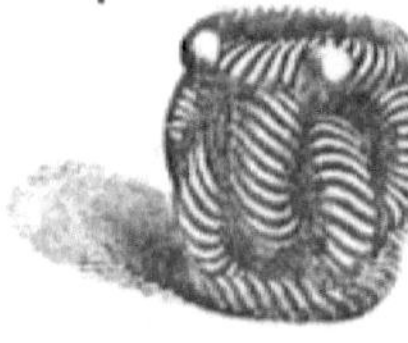

perishable nature. They are of various kinds and sizes, and are more or less ornamented. The accompanying

Fig. 296.   Fig. 297.

examples (figs. 294, 295, 296, and 297), will be sufficient to direct attention to these interesting relics. A number of beads, said to have been found with undoubted Roman remains, are shown on fig. 298.

Fig. 298.

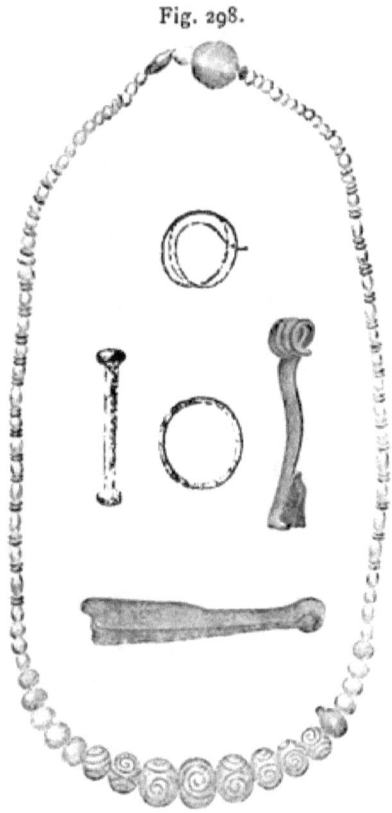

The Coins found along with Romano-British interments are, of course, of various emperors and of various periods. They are only occasionally found, and, when discovered, cannot, it must be remembered, be taken as any criterion as to date of deposit, or, indeed, cannot be considered alone as evidence of the barrow or interment belonging to the Romano-British period. The Romans seem to have sowed their coins broadcast over the whole length and breadth of the land, to have thrown them about as they would useless chaff, to have buried them in urns in every conceivable place, and to have deposited them, either singly or otherwise, in the barrows of their predecessors. It is unnecessary to speak, then, of the varieties of coins which are from time to time turned up by the antiquary in his researches into the early grave-mounds. They form but a thousandth part of the coins which are found away from interments.

It may, however, be well, as showing the relative proportions of the coins of different emperors found in this country, to give the following analysis, by Mr. Roach Smith, of more than eleven hundred coins picked up at different times in one locality—Richborough in Kent.

Augustus	7	Valerianus, junior	1
Agrippa	1	Galliense	19
Tiberius	2	Salonina	4
Antonia, wife of Drusus, sen.	1	Postumus	10
Caligula	2	Victorinus	14
Claudius	15	Marius	1
Nero	11	Tetricus	13
Vespasian	13	Claudius Gothicus	15
Titus	1	Luntillus	2
Domitian	10	Aurelianus	4
Nerva	1	Tacitus	5
Trajan	7	Florianus	1
Hadrian	5	Probus	7
Sabina	5	Carinus	1
Ællius Cæsar	1	Numerainus	2
Antoninus Pius	5	Diocletianus	8
Faustina	3	Maximianus	16
Marc Aurelius	4	Caräusius	94
Faustina	5	Allectus	45
Lucius Verus	2	Constantius	4
Lucil.a	1	Helena	8
Commodus	2	Theodora	13
Severus	5	Galerius Maximianus	1
Julia Domna	3	Maxentius	2
Caracalla	3	Romulus	1
Julia Maesa	1	Licinius	12
Severus Alexander	7	Licinius, junior	1
Gordianus	6	Constantine the Great	149
Philippus	4	Fausta	2
Valerianus	3	Crispus	18

Delmatius	1[2]	Valens 39
Constantine II.	98	Gratianus 49
Constans	77	Theodosius 14
Constantius II.	42	Magnus Maximus . . 6
Urbs Roma	52	Victor 3
Constantinoplis	60	Eugenius 1
Magnentius	21	Arcadius 27
Decentius	4	Honorius 8
Julianus II.	7	Constantine III. . . 1
Helena	1	
Jovianus	1	Total . 1144
Valentinianus	22	

Of these coins, fifty-six only were of silver, six of gold, fifteen of billon, or base silver, and the remainder were of brass, the greater portion being, naturally, what are denominated "third brass."

CHAPTER X.

Romano-British Period—Arms—Swords — Spears, etc.—Knives — Fibulæ—Armillæ—Torques of Gold, etc.—Other Personal Ornaments—Horse-shoes.

OF ARMS but few examples are found in grave-mounds, although more abundant in the neighbourhood of Roman stations and towns. They consist of swords, daggers, spear-heads, and other weapons. They are, however, perhaps the most scarce of any remains of the period. The swords of bronze (figs. 299 and 300) which have fre-

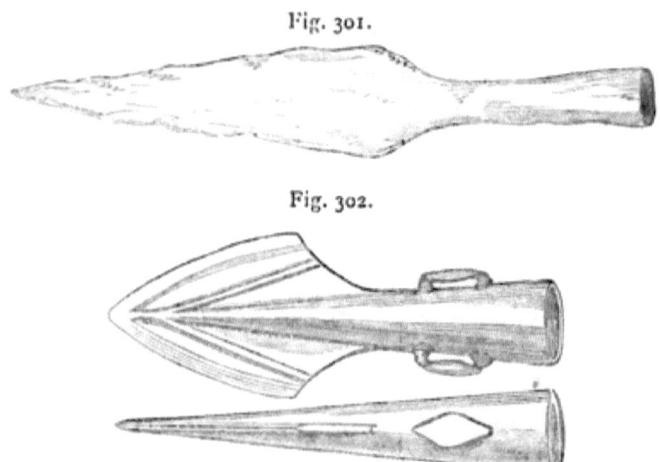

Fig. 301.

Fig. 302.

quently been ascribed to the British period, are now pretty generally admitted to belong possibly to Roman times. The examples engraved are of the most general type, as are also

SWORDS.   191

the next engravings of spear and lance heads. The first
(fig. 301), which is of iron, is from Little Chester, where it

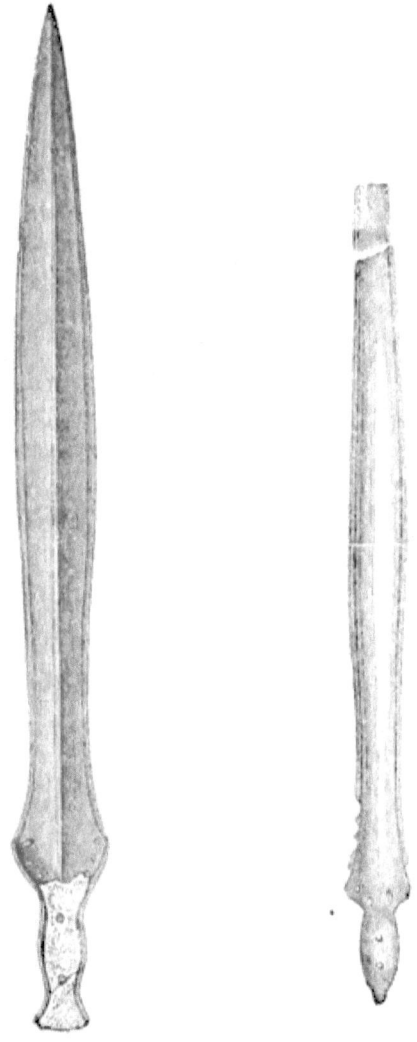

Fig. 299.   Fig. 300.

was found along with human remains. Fig. 302 is of

bronze, and is, as will be seen, of somewhat unusual form, and has a loop on either side. The next (fig. 303) is of

Fig. 303.

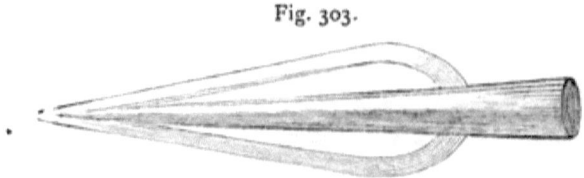

bronze, and is three and a half inches long. It is of remarkably good form, deeply socketed, like the preceding example, and of a kind of leaf shape. Arrow-heads are also occasionally found. Of these the example here engraved is a good type. It is of bronze, and measures about an inch and a quarter in length.

Fig. 304.

Iron knives are occasionally found with interments. Some remarkable instances of this have been recently brought to light near Plymouth, and others again at Wetton and other places. The knives are of the form engraved on fig. 305. They appear to have had wooden handles, which,

Fig. 305.

of course, except small traces of texture, have entirely decayed away. Another knife, although not actually found with an interment, shows the form so well that it is here engraved. It was nine and a half inches long, of peculiar

Fig. 306.

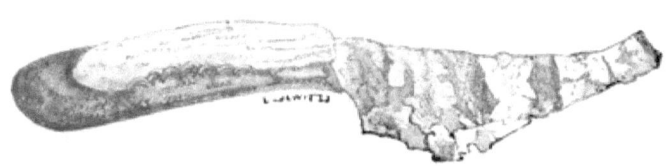

shape, still retaining its handle of stag's horn, rubbed or worn smooth; the good preservation of which we may attribute to having been imbedded in the fire-hardened earth, and sufficiently deep to secure it from injury by the fire. With the knives in the Plymouth cemetery were found portions of scissors, of the form of the sheep-shears of the present day, and these have also been found in other localities. They were of iron, and several fragments of other implements of the same material were at the same time discovered.

Of FIBULÆ an almost endless variety in form, in size, and in material has at one time or other been exhumed. They are, however, but very occasionally found with interments. The most usual form, perhaps, is that which is commonly called harp-shaped, or bowed, and this is of such extreme variety that scarcely two examples out of the hundreds that are known are precisely alike. Several have a cross bar at the top, and are hence called " cruciform " (figs. 307, 310 to 312, and 315). Others have coiled springs of wire at the top, variously fashioned. Some of these are extremely complicated and ingenious, as will be seen by the engraved examples. The more simple of the twisted springs, a coiled spring only, formed by the end of the bow being attenuated into the pin, is known as the " rat-trap spring," from its coiled resemblance to the spring used

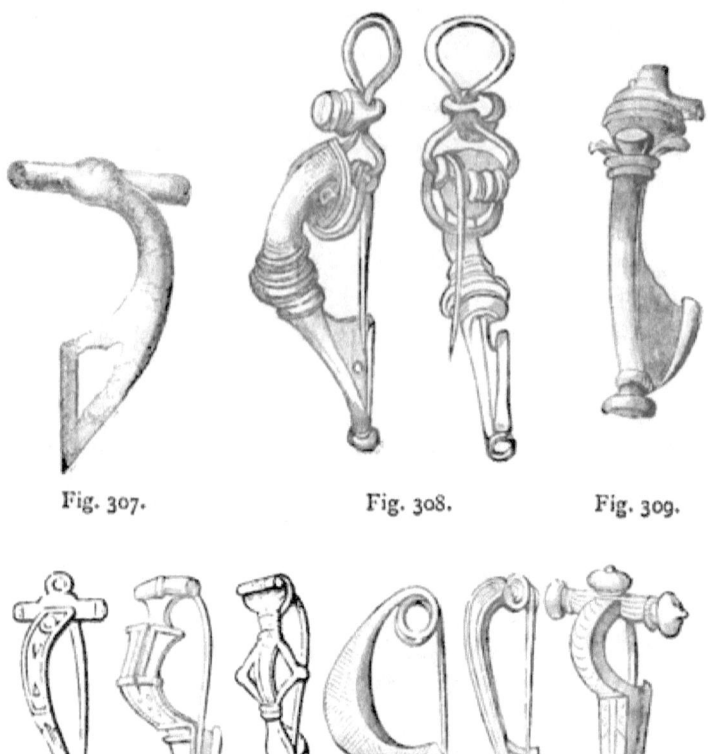

Fig. 307.   Fig. 308.   Fig. 309.

Fig. 310. Fig. 311. Fig. 312. Fig. 313. Fig. 314. Fig. 315.

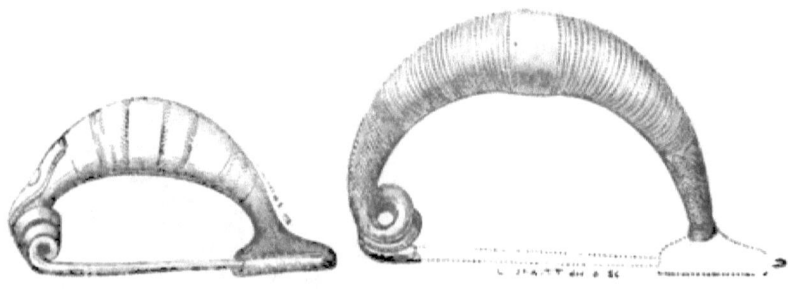

Fig. 316.   Fig. 317.

in those "vermin killers." Examples to show this form are here given (figs. 313, 314, 316, and 317). This form of fibula is generally known as the "dolphin" shape. Occasionally wire only, twisted in like manner as recently reproduced for skeleton shawl pins, are found. Sometimes the fibula really assumed the form of an animal, a bird or a serpent, with an inflated body. One of this character is

Fig. 318.

engraved on fig. 318. It is of one continuous piece of bronze, and the pin, having a coiled spring, answers to the tail of the serpent, and hooks into a projection on the neck.

The ornamentation is as varied as the form. Sometimes they are chased or engraved in minute patterns of rows of dots, scales, etc.; at others, enamelled or inlaid; and at others, again, raised ornaments are riveted upon their surface. Instances of S-shaped fibulæ also occur, as do many other grotesque forms.

Circular fibulæ are occasionally met with, and these,

Fig. 319.   Fig. 320.

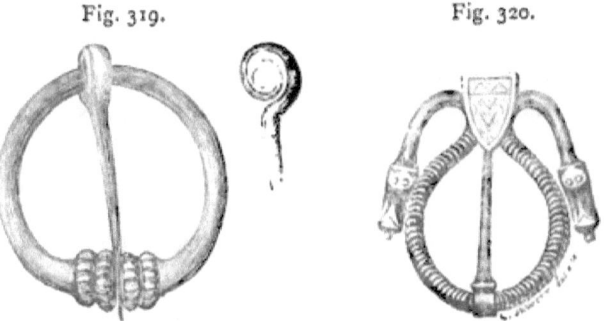

like the bowed forms, vary very considerably in design.

Sometimes they are flat on the face, and enamelled or inlaid in different colours. One of the most curious, but elegant, modifications of the circular form is fig. 320, where the ends, which are serpents' heads, are turned back to the sides of the body.

ARMILLÆ, or bracelets, are found both in bronze, in silver, and in gold. They vary very considerably in form. Of these, one example (fig. 321) will be sufficient. The pair

Fig. 321.

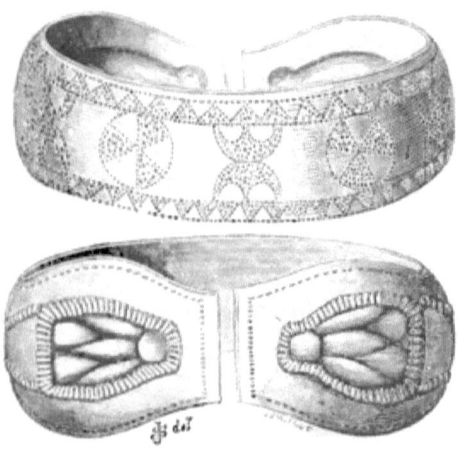

here represented are of base silver, and bear evidence of having been much worn. Examples of analogous type have been found at Castleford and other places. Other armlets partake more of the character of torques, torquis, or collar; and others, again, are simply bars of metal, twisted in one or more coils, like a spiral spring, around the wrist.

While speaking of armlets and torques, it may not be out of place, as I purposely omitted them in the Celtic division of this work, to say a few words about the latter. There can be no doubt that the torque was worn both by the ancient Briton and by his Roman conqueror, and therefore it is perhaps best, as it is at present not easy to say which of the known examples are to be attributed to the earlier and which to the later of these periods, to speak of them generally under this head.

The torque, or torquis, is said, by ancient writers, to have been first used by the Persians and by the nations of Northern and Western Europe. Virgil describes it as worn by the Trojans when they came to colonize Italy:—

> "Omnibus in morem tonsa coma pressa corona,
> Cornea bina ferunt præfixo hastilia ferro;
> Pars leves humero pharetras; it pectore summo
> Flexilis obtorti per collum circulus auri."

It is first mentioned in Roman history in the year 360 B.C., when Manlius, having torn a torque of gold from the neck of a vanquished Gaul—here is evidence of its being a decoration worn by a similar race to our ancient British population *before* being spoken of in Roman history—placed it on his own, and received, from this circumstance, the name of Torquatus. From this time the practice was adopted in the wars with the Gauls—the example set by Torquatus Manlius being frequently followed by the Roman leaders,—and the torque being adopted as a reward for military merit. "The Roman writers speak of them as worn by the Britons; and the Queen of the Iceni, Boadicea, is described by Dion Cassius as having a torquis of gold round her neck. This was the metal of which they were usually made. They consisted of a long piece of gold, twisted or spiral, doubled back in the form of a short hook at each end, and then turned into the form of a circle." The torque was known to, and worn by, the Egyptians, the Persians, Persepolitans, the Gauls, and the Britons, as well as, later on, to the Romans, and it was very usual, as is evident by the many examples which have been found,

198   GRAVE-MOUNDS AND THEIR CONTENTS.

with the Irish celts. The most usual forms will be found engraved in the catalogue of the Royal Irish Academy, the largest known example measuring five feet seven inches in length. A remarkably fine example of this type, found on the borders of Derbyshire and Staffordshire, measures three feet nine and a quarter inches. Many other varieties are found, sometimes formed of square bars of gold twisted spirally, sometimes of flat bars of the same metal twisted in a lighter manner, and sometimes, again, of more than one bar twisted together. The ends, too, are of various forms: sometimes being simply hooks, and at others swelling out into cup-shaped terminations, and at others partaking of the form of a serpent's head, etc. A very remarkable torque, now belonging to Her Majesty, was found in 1848 in Need-

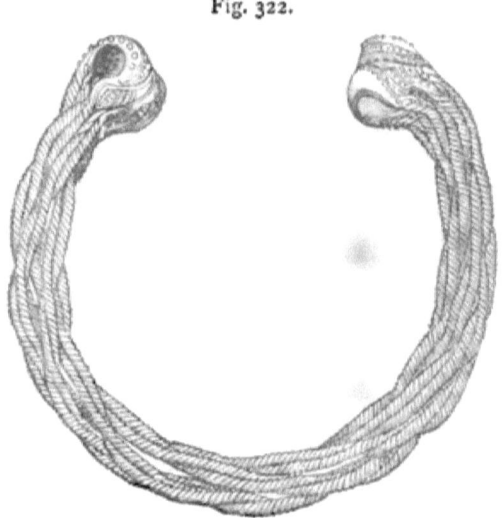

Fig. 322.

wood Forest, and is here engraved (fig. 322). It is formed of eight cords of gold plaited together, and weighs 1 lb. 1 oz.

7 dwts. and 10 grains. Another example of a different character, from Ireland, is here given (fig. 323).

Fig. 323.

Side View

It is safer, perhaps, although there is no doubt that torques were worn by the Romans, to assign them to the British period than to that of their conquerors. Much, however, necessarily depends on the remains found with them, and the locality where discovered.

Other personal ornaments, and bone and bronze pins, hairpins, etc., are occasionally found, but need no special notice here. Instruments of the toilette, too, are occasionally discovered. Prominent among these is the mirror, or speculum, which is sometimes found in the graves of Roman ladies. Among the most interesting of these are some found in a Roman cemetery at Plymouth. They consist of a circular plate of polished metal, generally of bronze set in a frame of the same metal or otherwise, and have a handle to hold them by. They are of much the same form as the small handled toilet glasses of the present day. The back was generally, as in the case of the Plymouth examples, "ornamented with a considerable quantity of scroll engraving. The pattern of one of these consists of three circular figures, the two bottom ones being larger than that which I take to be the central top one. Although each circular scroll differs from the others, they are evidently figured upon one general plan; the lines within, being segments of circles of various sizes, form crescents with various modifications. Some portions of the engraving, in order to give solidity to its character, were filled in with numerous striated spots, consisting of three lines one way and three lines at right angles to them. The entire surface of the mirror was surrounded by a narrow border or rim, which was formed of a separate piece, and folded over the margin. This specimen was damaged in many parts, particularly upon the under surface, and some of the edge was entirely eaten away, but where the rim was preserved the plate was not only in good preservation, but not even oxidized, retaining the bright colour of the bronze as perfectly as when, probably, in use by its ancient possessor. A second had the handle attached to it. The handle is cast in one piece in the form of a loop, having been made by folding one half back against the other, and securing them in that position by a band, the two free ends being spread out to hold the mirror, which is received in a groove, and supported on each side by a scroll work of bronze, of much of which, although lost, the impression still remains upon the plate. The greater diameter of the mirror is eight inches, that of the handle of the duplicate specimen, which is supposed to be of the same size as the missing handle, is four inches." Several of these mirrors have been found in the cemeteries at Colchester and in other places.

Combs, both of wood and bone, are also found in the interments, as are strigils, tweezers, locks and keys of numberless forms and sizes, remains of small caskets, and a great many other articles. Of combs I shall say a few words when speaking of those of the Anglo-Saxon period.

Horse-shoes of this period are occasionally met with in interments when the horse has been buried with his rider, or otherwise. One example, so as to show the form, will

Fig. 324.

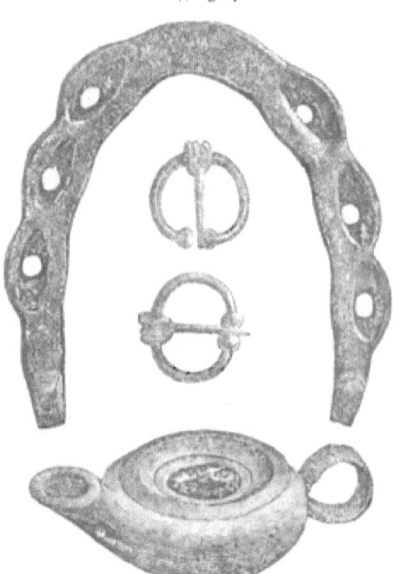

be sufficient. It was found at Gloucester some years ago, along with the lamp and circular fibulæ here engraved, and with other relics of the same period. Of the other articles it will not be necessary to make further mention.

CHAPTER XI.

Anglo-Saxon Period—Distribution of Anglo-Saxon Population over England—General characteristics of Grave-mounds—Modes of Burial—Poem of Beowulf—Interments by Cremation and by Inhumation—Articles deposited with the Dead—Positions of the Body—Double and other Interments—Burial in Urns—Cemeteries and Barrows.

THE grave-mounds and cemeteries of the Anglo-Saxon period present marked and decided features of difference to those of either of the preceding periods; and again, the characters of these mounds and cemeteries vary in different parts of the kingdom, according as such districts were inhabited by different tribes or peoples.

The date usually assigned to the first coming of the Saxons into England, after the final departure of the Romans, is the middle of the fifth century. They landed on the Isle of Thanet, and shortly afterwards established themselves in Kent, and became a kingdom. "Within thirty years another body of Saxons settled upon the south coast of Britain, taking possession of the tract now called Sussex, or the South Saxons. At the beginning of the sixth century a third detachment from the same Germanic family landed further westward, and founded the kingdom of the West Saxons, in which was included the Isle of Wight. From the same source which supplies the brief notices of these events, we learn that towards the middle of the sixth century were formed the states of the East and Middle Saxons, in the districts which, in consequence, took the names of Essex and Middlesex. We also gather that the Angles who settled in the east and north-east of Britain, and in the interior parts, probably made their first descents towards the middle of the sixth century; so that the kingdoms known as those of the East Angles (Norfolk, Suffolk, and Cambridgeshire), the Middle Angles, the Northumbrians (from the Humber northwards), and Mercia (on the borders of Wales), appear not to have been definitely settled until at least a century after the landing of the Saxons in Kent, in A.D. 449. Vague and unsatisfactory as are most of the details of Saxon history, the gradual subjugation of Britain by successive immigrations of Teutonic tribes may at least be accepted as the most reconcilable with reason; and there seems nothing very repugnant to the more rigid rules of criticism to regard these tribes under their historic designation of Jutes, Saxons, and Angles; and, further, to believe that at least a century was required to transform Britain, after the Romans, into a heptarchy of Teutonic kingdoms.

"Testing our Saxon antiquities with reference to the usually received chronology of the advent and settlement in Britain of the Teutonic tribes, it would be no unimportant result should they be in accordance with accepted historical facts. They will be invested with novel and higher interest if they should be found to carry in their form and character certain peculiarities which suggest earlier and later dates, and a diversity of parentage. For instance, if in the remains of the Kentish Saxons and in those of the Isle of Wight we may recognise, from close resemblance to each other, the weapons, the ornaments, and the domestic implements of the Jutes; if, in the cemeteries of Cambridgeshire, Suffolk, and Norfolk, we may, in like manner, identify the funeral usages of the Angles; and in remains found in the midland and western districts see still different peculiarities, but which point to a kindred origin; it is not improbable that discoveries may enable us to resuscitate, as it were, our remote predecessors; to restore to those of the various Saxon kingdoms the very objects which accompanied them when living: to the men, their weapons; to the women, their peculiar jewellery, and those more humble and homely objects which we may look upon as emblems of their domestic virtues. It is not a slight analogy in some instances only that will establish this theory; it must spring from the remains themselves, and be palpable and convincing, or it must be rejected."*

Bearing this in mind, and also bearing in mind the modifications which only a few years make in fashions and customs; and also bearing in mind that although for convenience sake, as well as for want of more definite knowledge, we call the whole population by the one term of Anglo-Saxons, yet they were divided into as distinct classes, or families, or tribes, as at the present day; we shall quite readily understand why the modes of burial, and the objects found in the graves, of one district are different from those, although coeval, found in others. At the present day we use the general term Englishmen for the whole of our population, and no better or clearer term could be adopted; but we must bear in mind that the differences both of appearance, of habits, of customs, of dialect, nay, of almost everything, are as marked among us as if the inhabitants of the various counties were each settlers from different nations. The men of Derbyshire, for instance, are as far removed as well can be in general character and in language from those of Somersetshire; and these, again, are both totally dissimilar from the "Men of Kent," from the Lancashire operative, from the Yorkshiremen, or the men of Devonshire, Hampshire, and many other counties. Each of these districts has, and always has had, and long, long may it continue to have! its own peculiar customs, its own peculiar habits, its own

* C. R. Smith.

peculiar observances; and each has what might almost be termed a *nationality* of its own, which it holds despite the levelling influence of railways and other modern contrivances. If it is so at the present day, with a settled population of so many centuries' standing, how much more must it have been so when each district was peopled by a different tribe of settlers, speaking to some extent different languages, holding different views, following different occupations, and observing different customs!

The grave-mounds and cemeteries of these different districts exhibit a marked difference in modes of burial, in style and decoration of pottery, and in characteristics of other remains, which will be made apparent in the following *resumé* of their varied contents. Thus, as Mr. Smith says, "in Kent one of the most conspicuous features in the Saxon sepulchral remains is the richly ornamented circular fibulæ. These are sparingly found beyond the district occupied by the earliest Saxon settlers. When they do occur, here and there, they are exceptions; but throughout the county of Kent it would be a rare occurrence to discover a Saxon funeral deposit without an example of this elegant and peculiar ornament. In Suffolk, in Norfolk, in Cambridgeshire, in Northamptonshire, in Leicestershire, and further north, these circular fibulæ do but casually appear, but others of a totally distinct character abound. In Berkshire, Oxfordshire, and Gloucestershire are found saucer-shaped fibulæ unlike either of these two classes, and forming a third variety. In Suffolk, in Cambridgeshire, in Leicestershire, and in other parts, have been repeatedly found metal implements or ornaments, which I have designated by the modern name of *chatellaine*, to give some notion of their form and use. These remarkable objects in no instance have been found in Kent, but other objects have been found in Kentish barrows which have nowhere else been discovered."

The sepulchral remains of the Anglo-Saxons are of two general classes—barrows and cemeteries—and in these the modes of burial have been both by inhumation and by cremation.

The grave-mounds, or barrows, are, as a rule, of much less altitude, and of smaller dimensions, generally, than those of either of the preceding periods. In some districts they are found in extensive groups, frequently occupying elevated sites; at other times they are solitary, and frequently the elevation above the surrounding surface is so slight as to be scarcely perceptible except to the most practised eye. Fortunately the mounds and cemeteries are particularly rich in remains, and thus enable us to form a clearer idea of the habits, and manners, and lives of our Saxon forefathers than we can of their predecessors. In Kent, Sussex, and the Isle of Wight, Saxon graves abound on the Downs; and in Derbyshire, Oxfordshire, Gloucestershire, Northamptonshire, Lincolnshire, Cambridgeshire, Suffolk, Norfolk, and Yorkshire, cemeteries of more or less extent and importance exist, with here and there a solitary barrow, or a group of barrows. Like their Roman predecessors the Anglo-Saxons, to some extent, took possession of, and buried in, the grave-mounds of the Ancient Britons, and it is not a very unusual occurrence to find overlying the primary deposit an interment of the Saxon period.

Fortunately an early Anglo-Saxon poem, recounting the adventures of the chieftain Beowulf, is preserved to us, and gives us a valuable and highly graphic and interesting description of the ceremonies attendant on his burial; the lighting of the funeral pyre, the burning of the body of the hero, the raising of the mound over his remains, and the articles placed beside him in his last home. Dying he

        bæð þæt ʒe ʒe-pophton
        æfteꞃa piueꞃ ðæðū
        in bæl-ꞃꞇeðe
        beoꞃh þone heán
        micelne and mæꞃne.—

Which is translated :—

> "he bad that ye should make,
> according to the deeds of your friend,
> on the place of the funeral pyle,
> the lofty barrow
> large and famous."

His request was carried out, the funeral pile raised, and every preparation befitting his deeds was made. The pile was—

> "hung round with helmets,
> with boards of war,*
> and with bright byrnies, †
> as he had requested.
> Then the heroes, weeping,
> laid down in the midst
> the famous chieftain,
> their dear lord.
> Then began on the hill,
> the warriors, to awake
> the mightiest of funeral fires;
> the wood-smoke rose aloft
> dark from the fire;
> noisily it went
> mingled with weeping."

The body of the hero having been consumed by the wood-fire, in the midst of weeping friends, the people began to raise the barrow over his ashes. This mound—

> "was high and broad,
> by the sailors over the waves
> to be seen afar.
> And they built up
> during ten days
> the beacon of the war-renowned.
> They surrounded it with a wall
> in the most honourable manner
> that wise men
> could desire.
> They put into the mound
> rings and bright gems,
> all such ornaments

* Shields.   † Armour.

>   as before from the hoard
>   the fierce-minded men
>   had taken;
>   they suffered the earth to hold
>   the treasure of warriors,
>   gold on the earth,
>   where it yet remains
>   as useless to men
>   as it was of old."

When the burial was simply by inhumation, the body appears usually to have been placed in a shallow grave, over which the mound was raised. The graves were of rectangular form, and of various depths. On the floor of the grave or pit the body was laid flat on its back, the arms straight down by its sides, the hands resting on the pelvis, and the feet close together. It was buried in full dress, and surrounded by a number of articles pertaining to the deceased—both personal ornaments, domestic instruments and vessels, and other things—and that had been used or valued by him or her. Sometimes the body was enclosed in a wooden chest or coffin before being placed in the grave. The grave, in either of these cases, was then filled in—usually with a tempered or "puddled" earth, which formed a close and extremely compact mass—and the mound raised over it. This mound or hillock was called a *hlæw*, or a *beorh, beorgh*, or *bearw*, from the first of which the name now commonly used, *low*, is derived, and from the last the equally common name *barrow* originates.

With the females, necklaces, rings, ear-rings, brooches, chatelaines, keys, buckets, caskets, beads, combs, pins, needles, bracelets, thread-boxes, tumblers, and a variety of other articles were found. With the males, swords, spears, knives, shields, buckles, brooches, querns, draught-men, etc., etc., are found. The warrior was usually laid, in his full dress, flat on his back (as already spoken of); his spears lying on his right side, his sword and knife on his left, and his shield laid on the centre of his body. The accompanying engraving (fig. 325) of a grave opened by the late Mr. Bateman, on Lapwing Hill, will pretty tolerably illustrate this mode of Anglo-Saxon burial. Beneath the bones of the skeleton were "traces of light-coloured hair, as if from a hide, resting upon a considerable quantity of decayed wood, indicating a plank of some thickness, or the bottom of a coffin. At the left of the body was a long

Fig. 325.

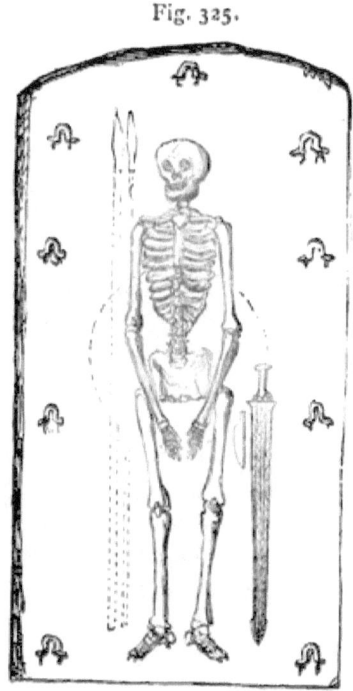

and broad iron sword, enclosed in a sheath made of thin wood covered with ornamental leather. Under or by the hilt of the sword was a short iron knife, and a little way above the right shoulder were two small javelin heads, four and a half inches long, of the same metal, which had lain so near each other as to become united by corrosion. Among the stones which filled the grave, and about a foot from the bottom, were many objects of corroded iron, including nine loops of hoop iron (as shown in the engraving) about an inch broad, which had been fixed to thick wood by long nails; eight staples, or eyes, which had been driven through a plank, and clenched; and one or two other objects of more uncertain application, all which were dispersed at intervals round the corpse throughout the length of the grave, and which may therefore have been attached to a bier or coffin in which the deceased was conveyed to the grave from some distant place. Indications existed of the shield having been placed in its usual position over the centre of the body, but no umbone was in this instance found. The mounds are usually, as in this instance, very low, frequently not being raised more than a foot above the natural surface of the ground. The earth was, as I have stated, usually "puddled" or tempered with water, and thus the body in the grave became closely imbedded in a compact and tenacious mass.

That the tempering, or puddling, was accompanied with some corrosive preparation, there can be little doubt; for it is a fact, though a very remarkable one, that whilst the skeletons of the Celtic period are found in good condition, and in some instances perfect and sound, those of the Anglo-Saxons have, almost invariably, entirely disappeared. Thus, in a Celtic barrow, the primary interment of that period may be found in perfect condition, while the secondary interment, that of the Anglo-Saxon, although some centuries later in date, and some three or four feet nearer the surface, will have decayed away and completely disappeared. Thus, in a barrow at Wyaston, which had been raised over the body of a Saxon lady, every indication of the body had disappeared, with the exception of the enamel coating of the teeth, while a splendid necklace of beads, a silver ring, silver ear-rings, and a silver brooch or fibula, remained *in situ* where the flesh and bones had once been. Another instance (to which I shall have occasion again to allude) which may be named, was the barrow at Benty Grange—a mound not more than two feet in elevation, but of considerable dimensions, and surrounded by a small fosse or trench, raised over the remains of a Saxon of high rank. In this mound, although a curious and unique helmet, the silver mountings of a leather drinking-cup, some highly interesting and beautiful enamelled ornaments, and other objects, as well as indications of the garments, remained, not a vestige of the body, with the exception of some of the hair, was to be seen. The lovely and delicate form of the female and the form of the stalwart warrior or noble had alike returned to their parent earth, leaving no trace behind, save the enamel of *her* teeth and traces of *his* hair alone, while the ornaments they wore and took pride in, and the surroundings of their stations, remained to tell their tale at this distant date. In a barrow at Tissington, in which the primary (Celtic) interment was perfect, the later Saxon one had entirely disappeared, while the sword and umbone of the shield remained as they had been placed.

The mode of interment with the funeral fire, as well as the raising of the barrow, is curiously illustrated by the opening of two Saxon graves at Winster. A large wood fire had, apparently, been made upon the natural surface of the ground. In this a part of the stones to be used for covering the body, and some of the weapons of the deceased, were burned. After the fire was exhausted the body was laid on the spot where it had been kindled, the spear, sword, or what not, placed about it, and the stones which had been burnt piled over it. The soil was then heaped up to the required height to form the mound.

Usually, of course, the graves contain only one body, but instances occasionally occur in which two or more bodies have been buried at the same time. For instance, at Ozengal a grave has been opened which was found to contain two skeletons. They were those of a man and a woman who were laid close together, side by side, with their faces to each other. In another were three skeletons, those of a man, a woman, and a little girl. The lady lay in the middle, her husband on her right hand, and their little daughter on her left; they lay arm in arm. In other cases two or more interments have been found, usually lying side by side, on their backs.

In many Anglo-Saxon barrows, bones, thrown in indiscriminate heaps or otherwise, are found at the top, over the original interments. These are, very plausibly, conjectured to be the remains of slaves or captives slain at the funeral, and thrown on the graves of their master or mistress.

When the burial has been by cremation, the ashes, after the burning of the body which is so graphically described in the extract I have given from Beowulf, were collected together and placed in urns. These were usually buried in small graves, and their mouths not unfrequently covered with flat stones. Some very extensive cemeteries where the burials have been by this mode, have been discovered in Derbyshire, Nottinghamshire, Northamptonshire, and other counties. With these it is very unusual to find any remains of personal ornaments or weapons. Two extensive and remarkable cemeteries of this kind have been discovered at Kingston and at King's Newton, both near Derby. At the first of these places an extensive cemetery was uncovered in 1844, and resulted in the exhumation of a large number of urns—indeed, so large a number that, unfortunately, at least two hundred were totally destroyed by the workmen before the fact of the discovery became known. On the surface no indication of burials existed; but as the ground had, some sixty years before, for a long period been under plough cultivation, and as the mounds would originally have been very low, this is not remarkable. The urns had been placed on the ground in shallow pits or trenches. They were filled with burnt bones, and the mouth of each had been covered with a flat stone. They were, when found, close to the surface, so that the mounds could only have been slightly elevated when first formed. Of the form of the urns I shall have to speak later on. The cemetery at King's Newton was discovered during the autumn of 1867, and a large number of fragmentary urns were exhumed. The mode of interment was precisely similar to that at Kingston, and the urns were of the same character as those there discovered. There were no traces, in either instance, of mounds having been raised, although most probably they had originally existed. To the pottery found in these cemeteries I shall refer later on. Cremation was the predominating practice among the Angles, including Mercia, and the modes of burning the body, and of interment of the calcined bones in ornamental urns, which I have described in the two cemeteries just spoken of, are characteristic of that kingdom. King's Newton is within three miles of Repton (Repandune), the capital of the kingdom of Mercia, and the burial place of Mercian kings, and Kingston is also but a few miles distant.

In some cases the burial has been without urns—the ashes being simply gathered together in a small heap in the grave, or on the surface, and the mound raised over it.

I will now, as in the previous divisions, proceed to speak of the more usual descriptions of relics which are found in the grave-mounds of the Anglo-Saxons, and I will, as in those divisions, commence with the fictile remains.

CHAPTER XII.

Anglo-Saxon Period — Pottery, general characteristics of — Cinerary Urns — Saxon Urn with Roman Inscription — Frankish and other Urns — Cemeteries at King's Newton, etc. — Mode of Manufacture — Impressed Ornaments.

THE pottery of the Anglo-Saxon period, so far as examples have come down to us, are almost, if not entirely, confined to sepulchral urns. We know, from the illuminated MSS. of the period, to which we are accustomed to turn for information upon almost any point, that other vessels — pitchers, dishes, etc. — were made and used, but for those which have come down to us we are indebted to the grave-mounds; and, in these, sepulchral vessels, almost exclusively, are found to occur. Cinerary urns are, therefore, almost the only known productions of the Saxon potteries, and these, like those of the Celtic period, were doubtless, in most cases, made near the spot where the burial took place, and were formed of the clays of the neighbourhood. This is proved, incontestably, in the case of the urns found at King's Newton, where the bed of clay still exists, and has very recently been used for common pottery purposes.

The shapes of the cinerary urns are somewhat peculiar, and partake largely of the Frankish form. Instead of being wide at the mouth, like the Celtic urns, they are contracted, and have a kind of neck instead of the overhanging lip or rim which characterizes so much of the sepulchral pottery of that period. The urns are formed by hand, not on the wheel, like so many of the Romano-British period, and they are, as a rule, perhaps, more firmly fired than the Celtic ones. They are usually of a dark-coloured clay, sometimes nearly black, at other times they are dark brown, and occasionally of a slate or greenish tint, produced by surface colouring. The general form of these interesting fictile vessels will be best understood by reference to the engravings which follow. One of these (on fig. 326) will be seen to have projecting knobs or bosses,

which have been formed by simply pressing out the pliant clay from the inside with the hand. In other examples these raised bosses take the form of ribs gradually swelling out from the bottom, till, at the top they expand into semi-egg-shaped protuberances. The ornamentation on the urns from these cemeteries usually consists of encircling incised lines in bands or otherwise, and vertical or zigzag lines arranged in a variety of ways, and not unfrequently the knobs or protuberances of which I have just spoken. Sometimes, also, they present evident attempts at imitation of the Roman egg-and-tongue ornament. The marked features of the pottery of this period are the frequency of small punctured or impressed ornaments, which are introduced along with the lines or bands with very good effect. These ornaments were evidently produced by the end of a stick cut and notched across in different directions so as to produce crosses and other patterns. In some districts—especially in the East Angles—these vessels are ornamented with simple patterns painted upon their surface in white; but so far as my knowledge goes, no example of this kind of decoration has been found in the Mercian cemeteries.

Of these urns—the East Anglian, etc.—Mr. Wright (to whom, and to Mr. Roach Smith, is mainly due the credit of having correctly appropriated them to the Anglo-Saxon period), thus speaks :—

"The pottery is usually made of a rather dark clay, coloured outside brown or dark slate colour, which has sometimes a tint of green, and is sometimes black. These urns appear often to have been made with the hand, without the employment of the lathe; the texture of the clay is rather coarse, and they are rarely well baked. The favourite ornaments are bands of parallel lines encircling the vessel, or vertical and zigzags, sometimes arranged in small bands, and sometimes on a larger scale covering half the elevation of the urn; and in this latter case the spaces are filled up with small circles and crosses, and other marks, stamped or painted in white. Other ornaments are met with, some of which are evidently unskilful attempts at imitating the well-known egg-and-tongue and other ornaments of the Roman Samian ware, which, from the specimens, and even fragments, found in their graves, appear to have been much admired and valued by the Anglo-Saxons. But a still more characteristic peculiarity of the pottery of the Anglo-Saxon burial urns consists in raised knobs or bosses, arranged symmetrically round them, and sometimes forming a sort of ribs, while in the ruder examples they become mere round lumps, or even present only a slight swelling of the surface of the vessel.

"That these vessels belong to the early Anglo-Saxon period is proved beyond any doubt by the various objects, such as arms, personal ornaments, etc., which are found with them, and they present evident imitations both of Roman forms and of Roman ornamentation. But one of

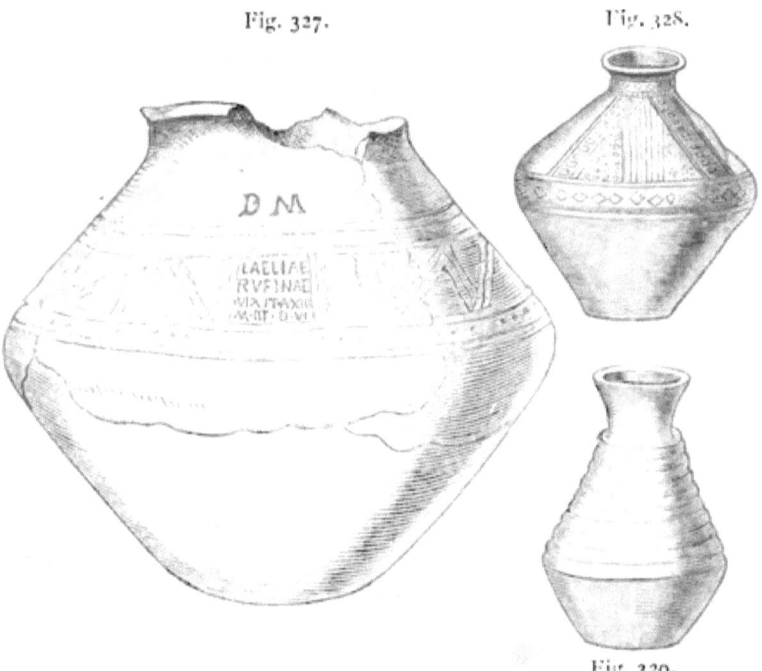

Fig. 327.   Fig. 328.

Fig. 329.

these urns has been found accompanied with remarkable circumstances, which not only show its relative date, but illustrate a fact in the ethnological history of this early period. Among the Faussett collection of Anglo-Saxon antiquities is an urn which Bryan Faussett appears to have obtained from North Elmham, in Norfolk, and which contained the bones of a child. It is represented in the accompanying engraving (fig. 327), and will be seen at once to be perfectly identical in character with the East Anglian sepulchral urns. But Mr. Roach Smith, in examining the various objects in the Faussett collection, preparatory to his edition of Bryan Faussett's '*Inventorum Sepulchrale*,' discovered on one side of this urn a Roman sepulchral inscription, which is easily read as follows :—

D. M.	' To the gods of the shades.
LAELIAE	To Lælia
RVFINAE	Rufina.
VIXIT·A·XIII	She lived thirteen years,
M·III·D·VI.	three months, and six days.'

To this Roman girl, with a purely Roman name, belonged, no doubt, the few bones which were found in the Anglo-Saxon burial urn when Bryan Faussett received it, and this circumstance illustrates several important as well as interesting questions relating to our early history. It proves, in the first place, what no judicious historian now doubts, that the Roman population remained in the island after the withdrawal of the Roman power, and mixed with the Anglo-Saxon conquerors ; that they continued to retain for some time at least their old manners and language, and even their Paganism and their burial ceremonies, for this is the purely Roman form of sepulchral inscriptions ; and that, with their own ceremonies, they buried in the common cemetery of the new Anglo-Saxon possessors of the land, for this urn was found in an Anglo-Saxon burial ground. This last circumstance had already been suspected by antiquaries, for traces of Roman interment in the well-known Roman leaden coffins had been found in the Anglo-Saxon cemetery at Ozingell, in the Isle of Thanet; and other similar discoveries have, I believe, been made elsewhere. The fact of this Roman inscription on an Anglo-Saxon burial urn, found immediately in the district of the Anglo-Saxon cemeteries, which have produced so many of these East Anglian urns, proves further that these urns belong to a period following immediately upon the close of what we call the Roman period."

The sepulchral vases found in the district of the middle Angles vary but slightly in form from the East Anglian burial urns. An example is given in fig. 328, from Chestersovers, in Warwickshire, where it was found with an iron sword, a spear-head, and other articles of Anglo-Saxon character.

"If we had not abundant proofs of the Anglo-Saxon character of this pottery at home," continues Mr. Wright, "we should find sufficient evidences of it among the remains of the kindred tribes on the Continent, the old Germans, or Alemanni, and the Franks. Some years ago an early cemetery, belonging to the Germans, or Alemanni, who then occupied the banks of the Upper Rhine, was discovered near a hamlet called Selzen, on the northern bank of that river, not far above Mayence, and the rather numerous objects found in it are, I believe, preserved in the Mayence Museum. They were communicated to the public by the brothers Lindenschmit, in a well-illustrated volume published in 1848, under the title '*Das Germanische Todtenlager bei Selzen in der Provinz Reinhessen.*' When this book appeared in England, our antiquaries were astonished to find in the objects discovered in the Alemannic cemeteries of the country bordering on the Rhine a character entirely identical with that of their own Anglo-Saxon antiquities, by which the close affinity of the two races was strikingly illustrated. More recently, the subject has been further illustrated in the description by Ludwig Lindenschmit of the collection of the national antiquities in the Ducal Museum of Hohenzollern, and in other publications. About the same time with the first labours of the Lindenschmits, a French antiquary, Dr. Rigollot, was calling attention in France to similar discoveries in the cemeteries which the Teutonic invaders of Picardy had left behind them, and in which he recognized the same character as that displayed by the similar remains of the Anglo-Saxons in our island. Similar discoveries have been made in Burgundy and in Switzerland, the ancient country of the Helvetii; and it is hardly necessary here to do more than mention the great and valuable researches carried on by the Abbé Cochet among the Frankish graves in Normandy. It has thus become an established fact that the varied remains of the tribes, all of Teutonic descent, who settled on the borders of the Roman empire along the whole extent of the country from Great Britain to Switzerland, present the same character and bear a close resemblance.

A few figures will be sufficient to illustrate this resemblance as far as regards the pottery, and these are here given, in which figs. 330 and 332 are Alemannic vases from the cemetery of Selzen. It will be seen that they resemble exactly in form those East Anglian urns we have given in our plate, and the same ornamentation is also found among our Anglo-Saxon pottery. These urns are described as being usually made of the clay of the neighbourhood, in most cases turned on a lathe, but many of them imperfectly baked. They are found in graves where the body had not undergone cremation, and were used for containing articles of a miscellaneous description. In one grave, at the feet of the skeleton of a gigantic warrior, was found one of these urns, containing two bronze fibulæ, a comb, a number of beads, a pair of shears, flints and steel, and a bronze ring. Fig. 334 is an urn procured at Cologne, and is slate-coloured, with an ornament of circular stamps.

Figs. 331 and 333 are Frankish urns obtained by the Abbé Cochet from Londinières in Normandy, and show at a glance the identity of the Frankish pottery with the Germanic as well as with the Anglo-Saxon. The first of these is surrounded with a row of the well-known bosses, which are equally characteristic of the three divisions of this Teutonic

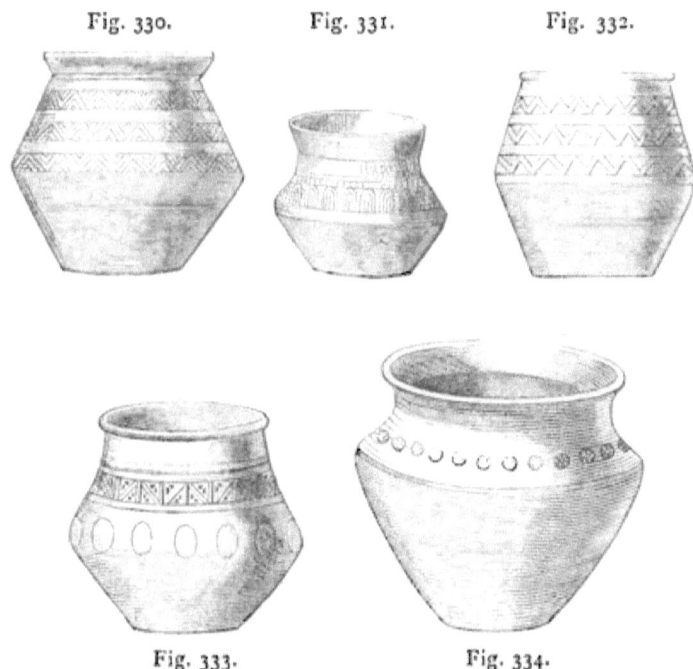

Fig. 330.   Fig. 331.   Fig. 332.

Fig. 333.   Fig. 334.

pottery, Anglo-Saxon, Frankish, and Alemannic. Above these bosses is an ornament identical with that of the East Anglian urn with the sepulchral inscription, given on fig. 327. The urn represented in fig. 331 has an ornament which is evidently an imitation of the egg-and-tongue ornament so common on the Roman pottery. The Abbé Cochet collected in the course of his excavations in Normandy several hundreds of these Frankish urns, which all present the same general character.

The next four examples are earthen vessels found in the lacustrine habitations of Switzerland, of which so much has been written during the last few years. Figs. 335, 336, and 337, are taken from the plates illustrative of the communications of Dr. Ferdinand Keller to the *Transactions* of the Antiquarian Society of Zurich, and fig. 338, also from the Zurich *Transactions*, and found in a Pfahlbau, near Allensbach on the Untersee, on the borders of Switzerland and Germany. A single glance will show a great similarity of form with those of the Anglo-Saxons from our own country.

The following engravings will exhibit a striking variety

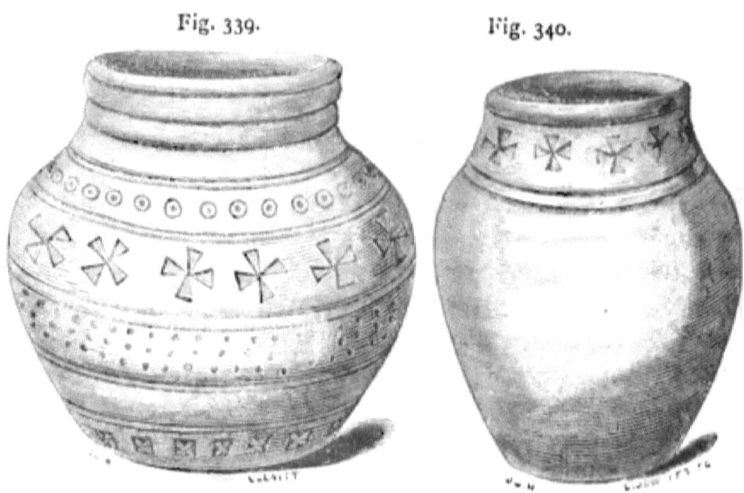

Fig. 339.   Fig. 340.

of cinerary urns of the Anglo-Saxon period, from the Mercian cemetery at King's Newton. Fig. 339 is six and a quarter inches in height. It is ornamented with encircling bands or lines and impressed ornaments. In the

Figs. 335 to 338.

upper band is a series of small circular indentations, with a dot in the centre of each, and in the lower band are three rows of dots. Between these bands is a series of indented

Fig. 341.   Fig. 342.

Fig. 343.

crosses, which may be described as in some degree approaching to crosses *patée* in form. At the bottom are also small square indentations, with diagonal lines. Fig. 342 is seven inches in height. It is ornamented with encircling

lines, the central band bearing a double row of dots; the
band at the bottom of the neck a series of small indented

Fig. 344.   Fig. 345.

quatre-foil flowers; and the lower one a series of square indentations with diagonal lines.   Fig. 341 is one of the

Fig. 346.

most elaborately ornamented urns which has ever been discovered.*   The remainder of the examples vary from these

* A detailed account of this discovery will be found, from the pens of
Mr. Briggs, the Editor, and others, in the "Reliquary," vol. ix.

15

and from each other, in point both of form and decoration. Some of these have herring-bone lines, others simple punctures, and others, again, encircling lines only. The marked

Fig. 347.

Fig. 348.

features of the pottery of this period is the frequency of small punctured or impressed ornaments, which are introduced along with the lines or bands, with very good effect.

ORNAMENTATION OF POTTERY.   227

These ornaments were evidently produced by the end of a stick, cut and notched across in different directions, so as to produce crosses and other patterns, and by twisted slips of metal, etc. In the annexed woodcut I have endeavoured to show two of the notched stick "punches," such as I have reason to believe were used for pressing into the soft clay,

Fig. 349.   Fig. 350.

and also two of the impressed patterns produced by it.

Fig. 351.   Fig. 352.

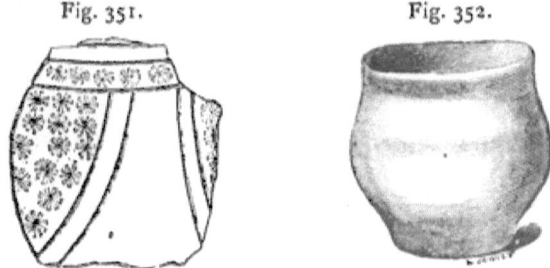

Other varieties of pottery found in the Anglo-Saxon graves are a species of cup, and upright vessels, one of which is shown on fig. 327. Fragments of pitchers have also occasionally been found, as also have portions of coarse dishes. In the Kentish graves, most of the pottery is of the Roman period, and consists of Samian pateræ and other vessels of that manufacture; and cups, etc., of the Upchurch and Castor wares, etc.

CHAPTER XIII.

Anglo-Saxon Period—Glass Vessels—Drinking-glasses—Tumblers—
Ale-glasses—Beads—Necklaces—Ear-rings—Coins, etc.

THE glass vessels found in the grave-mounds of the Anglo-Saxon period are principally drinking-cups of different forms, and decanter-shaped vessels, which are closely analogous in shape to our common glass toilet water-bottles. The Anglo-Saxons are supposed by most writers to have derived their knowledge of the art of glass-making from their Roman predecessors, but of this more proof is wanting. So very different in most of its characteristics is the Saxon glass from the Roman, that it is difficult to believe that the one is but an imitation of the other. The forms are in many instances similar to those found in Frankish graves, and it is certain that the art was practised simultaneously in the Saxon period in Germany, France, and our own country. The drinking-cups of glass were formed either rounded or pointed at the bottom, so that they could not stand, and thus when filled the liquor was obliged to be drunk off before the cup could be set down inverted on the table. From this circumstance our modern name for drinking-glasses—*tumblers*—takes its origin, although not now in the original sense, our present "tumblers" being particularly safe and firm when set on the table, and not necessitating the whole of the contents being quaffed at once. Figs. 353 and 354 exhibit two drinking-glasses of this kind, the first of which is ribbed. They are from the Kentish graves. Fig. 355 is a glass cup from Cow-Low, Derbyshire, found by the late

Mr. Bateman, and which, from the care which had been taken in enclosing it in a wooden box, must have been no little prized by the deceased lady. The cup, of thick green

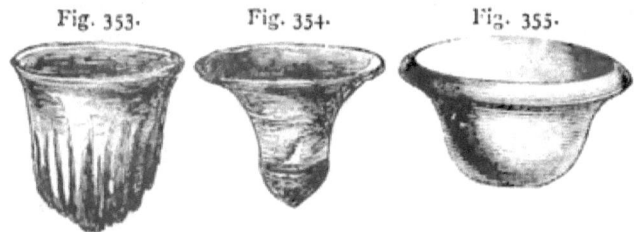

Fig. 353.   Fig. 354.   Fig. 355.

glass, a bone comb, some small instruments of iron, a piece of perforated bone, and a necklace with pendant ornaments, with other articles, were found enclosed in a box, or casket, made of ash wood, half an inch in thickness, with two hinges and a small lock, which had, when placed in the grave, been carefully wrapped in woollen cloth. The interment was in many respects a highly interesting one.

Fig. 356.   Fig. 357.

Fig. 358.

A cup of similar form is shown on fig. 358, and other examples of glasses are shown on the same group.

These examples, it will be seen, must have been held in the open palms of the hand, as is seen so frequently represented in illuminations, and must have been emptied of their contents before being returned, inverted, to the board.

Another form, figs. 356 and 357, is the long ale-glass, the shape of which is probably derived from the drinking-horns which were in use. They, and other of the Saxon glasses, were often ornamented with a raised thread or band on their outer surface, arranged either spirally or otherwise. In *Beowulf* these glasses are spoken of—

>Þegn nytte beheolð
>þe þeon handa bær
>hpoden ealʒ-pæʒe.

>"The Thane observed his office,
>he that in his hand bare
>the twisted ale-cup."

This form of glass is well illustrated in the next engraving, fig. 359, from a MS. of the twelfth century. In it the cup-bearer holds the glass in one hand and the jug in the other, from which he has just filled it. As an accompaniment to

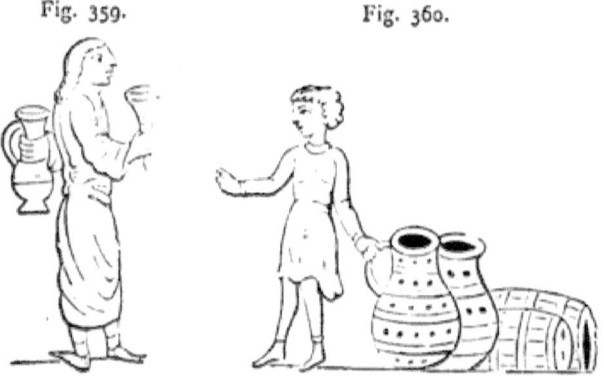

Fig. 359.   Fig. 360.

this I give another engraving, which shows the cellarer with the barrels and two large earthenware pitchers, which, it will be observed, are ornamented in precisely the same manner as some of the urns I have engraved. Another excellent example of the use of these glasses at a banquet

Fig. 361.

is shown on fig. 361, where a mixed company of males and females are seated at a banquet, and pledging each other in them. It is from the Cottonian MSS. in the British Museum.

Fig. 362.   Fig. 363.

The next two figures (362 and 363) show two of the decanter-shaped vessels to which I have alluded, and figs.

364, 365, and 366 again exhibit a different variety—one in which the ornament is formed of a number of what may almost be called handles—hollow protuberances, or claws,

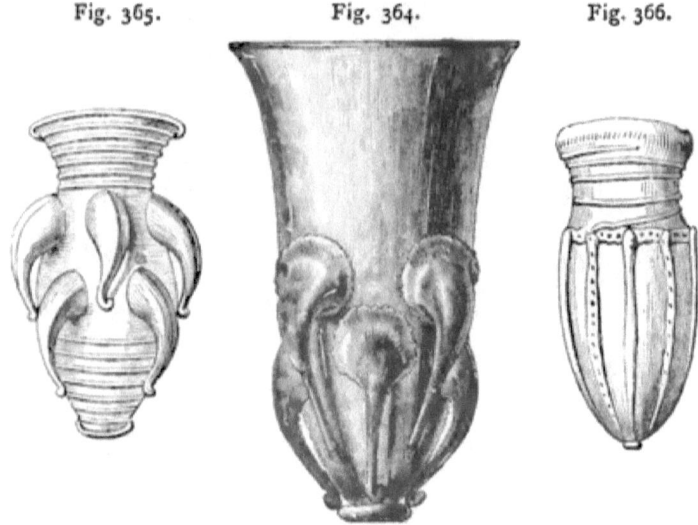

Fig. 365.   Fig. 364.   Fig. 366.

attached at the upper and lower ends. Many specimens of these curious glasses have been found in graves in different districts.

Among the most profuse of Anglo-Saxon remains are the beads and necklaces of glass, of amber, and of other materials, many of which are of extreme beauty. The greater part of the beads which are found are composed of glass, transparent and opaque; variegated clays of different colours; and of amber. Less frequently beads of amethystine quartz, of crystal, and of other rare natural substances are found. Sometimes the beads are formed singly, and at other times they are in couplets or triplets. Beads of metal—gold and silver—and of stones set in the same precious metals, have also been exhumed. Beads mounted on rings, or, more properly speaking, threaded on rings,

BEADS AND NECKLACES. 233

are of not unfrequent occurrence, and appear, in many instances, to have been intended for the ears. The three engravings (figs. 367, 368, and 369) will serve as examples

Fig. 367.

Fig. 368.   Fig. 369.

of beads. The first, engraved full size, is of glass, and is ornamented with red, white, and yellow waves. The other two are of clay, with yellow stripes. They are from Siberts-wold. The beads from the Kentish barrows are perhaps the most extensive in number, as well as the most varied in form, material, and ornamentation, of any. The next illustration (fig. 370) shows a series of twenty-seven beads, which formed the necklace of an Anglo-Saxon lady, whose grave was opened by Mr. Bateman at Wyaston. In this barrow, which was thirty-three feet in diameter, and four feet high in the centre, were discovered the remains of a human skeleton, consisting merely of the enamel crowns of the teeth, which, though themselves but scanty mementoes of female loveliness, were accompanied by several articles indicating that the deceased was not unaccustomed to add the ornaments of dress to the charms of nature. These comprise a handsome necklace of twenty-seven beads, a silver finger-ring, silver ear-rings, and a circular brooch or

Fig. 370.

fibula. Five of the beads are of amber, carefully rounded into a globular shape, the largest an inch diameter; the remaining twenty-two (two of which are broken) are mostly small, and made of porcelain or opaque glass, very prettily variegated with blue, yellow, or red, on a white or red ground. The finger-ring is made of thick silver wire, twisted into an ornamental knot at the junction of the ends. The ear-rings are too slight and fragmentary for description. The fibula is a circular ring, ribbed on the front, an inch and a half diameter, composed of a doubtful substance. The remains of the teeth show the person to have been rather youthful, and afford another instance of the extreme decay of the skeleton usual in Saxon deposits in this part of the country, whilst those which we have reason to reckon centuries more ancient are mostly well preserved. Rings of silver, with cylindrical, or globular, or flattened beads attached, are of common occurrence in the Kentish and other graves. Of pendants I shall speak a little later on.

Fig. 371.

Coins have only occasionally been discovered with Anglo-Saxon interments, and these have, in most instances, been of the preceding Roman period. Byzantine, Frankish, and Merovingian coins have likewise been found in the graves. Coins, to which loops are attached, so as to be worn as personal ornaments, are also found.

CHAPTER XIV.

Anglo-Saxon Period—Arms—Swords—Knives—Spears—Shields—Umbones of Shields—Buckles—Helmets—Benty-Grange Tumulus—The Sacred Boar—Grave at Barlaston—Enamelled Discs and pendant Ornaments, etc.—Horse Shoes.

THE arms of the Anglo-Saxons, so far as is known from the contents of their graves, consisted of swords, spears, knives, shields, daggers, etc., and occasionally with the men, besides these things, are found remains of helmets, ornaments from horse-trappings, buckles, axes, and many other articles.

The swords are straight-bladed, usually double-edged, with hilts of metal or wood. The scabbards were sometimes of wood, sometimes of leather, and sometimes again of bronze, and are often elaborately ornamented at the chape. The sword here engraved (fig. 372) was found in a barrow at Tissington, in Derbyshire. It had originally been enclosed in a wooden scabbard or sheath, which had apparently been covered with leather, and mounted with ornamented silver. Most of this ornamentation was decayed and lost, but sufficient remained to show that the sword had been of no ordinary beauty and value, and must have belonged to some person of note. The traces of silver ornamentation at the head are indicated on the engraving. The chape, which is simply rounded, is of silver, and the rivets still remain, as do also those by which the leather was attached to the wood. The sword is thirty-four inches in length,

Fig. 372.   Fig. 373.    Fig. 374.

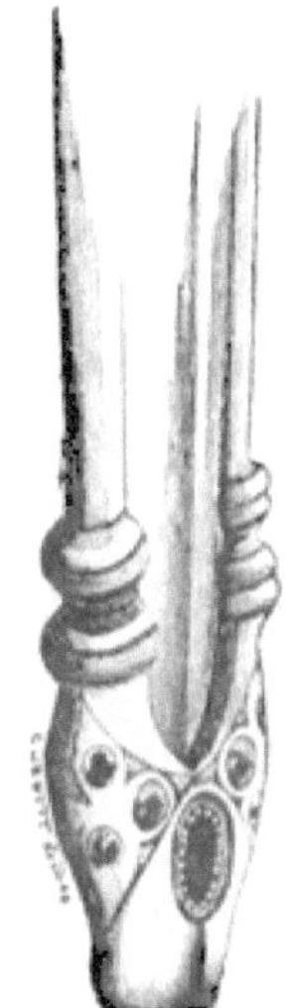

Fig. 375.

and two inches and a half in breadth. Across its upper part lay a small fragment of the shield, and near it, spread about, were a few pieces of iron, some of which, when joined together, proved to be a spear-head of the usual form of the period. It had doubtless been broken and disturbed at the time when the bones were dispersed by the planting of the trees.

A remarkably fine sword (fig. 373) was found in 1868 at Grimsthorpe. It is of iron, and remains encased in its bronze scabbard in a more perfect state than usual. The extreme length of the sword and scabbard, from pommel to chape, is thirty-one inches; the length of the scabbard from guard to point of chape, twenty-four inches. The breadth at the mouth is one inch and seven-eighths. The guard is of bronze, and is engraved on fig. 374. The scabbard is formed of thin plate bronze, and has an encircling band of the same material to hold the upper points of the chape to its sides. The length of the chape from the band is six inches and a half. The chape, which is exquisitely formed, is engraved on fig. 375, and will be seen to be of unusual beauty. It is in a remarkably perfect condition, and, being formed of bronze (the scabbards of the period to which it belongs being usually of wood with metal chape and fittings), is of great rarity and interest. The chape had been set with six small, and one large, stones, as will be seen by the engraving. Some of these, which were probably garnets, were remaining. They had been affixed to their places by small rivets passing through their centres. A series of fifteen examples of Anglo-Saxon swords (figs. 376 to 390) from illuminated MSS., etc., are here given for purposes of comparison. Some of these will be found to be of precisely similar form to those already given, and others, again, have trefoiled pommels.

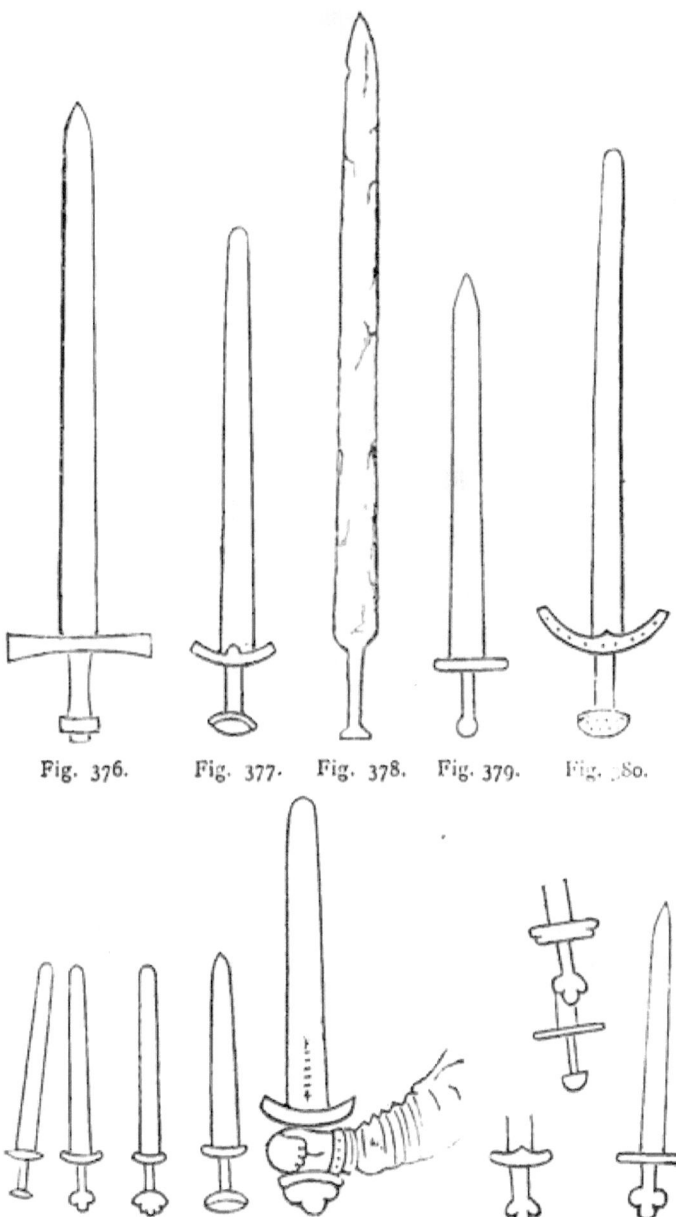

Fig. 376. Fig. 377. Fig. 378. Fig. 379. Fig. 380.

Figs. 381 to 390.

A good figure of a swordsman, with sword and shield, is also given on the next figure (fig. 391).

Fig. 391.

Swords with ornamental pommels and hilts are of rare occurrence, but examples occur in the Faussett and other collections. Probably those with ornamented hilts would also have their chapes correspondingly ornamented. In *Beownlf* occur these lines :—

> "When he did off from himself
> his iron coat of mail,
> the helmet from his head,
> gave his ornamented sword,
> the costliest of irons,
> to his servant."

And again :—

        anð þa hilt ȝomoð
        ſince fáȝe.

        (" And with it the hilt
          variegated with treasure.")

A remarkable hilt, bearing an inscription in Runic characters, was found at Ash, in Kent. It is of silver. On one side is the Runic inscription engraved in the metal, on the other a zigzag and other ornaments. A hilt of this kind must undoubtedly have been the one so graphically described in *Beowulf*, where a sword, inscribed with the name of its first owner and with other matters of extreme interest, is "looked upon" and pondered over. The passage is thus :—

        " He looked upon the hilt,
          the old legacy,
          on which was written the origin
          of the ancient contest ;
          after the flood slew,
          the pouring ocean,
          the race of giants;
          daringly they behaved ;
          that was a strange race
          to the eternal Lord,
          therefore to them their last reward
          through floods of water
          the ruler gave.
          So was on the surface
          of the bright gold
          with Runic letters
          rightly marked,
          set and said,
          for whom that sword,
          the costliest of irons,
          was first made,
          with twisted hilt and variegated like a snake."

The runes on the hilt first spoken of and engraved

would doubtless, if properly translated, tell as pleasant and as interesting a story as the one narrated by Beowulf.

The knife or dagger (the *seax*), which is of iron, is of

Figs. 392 to 396.

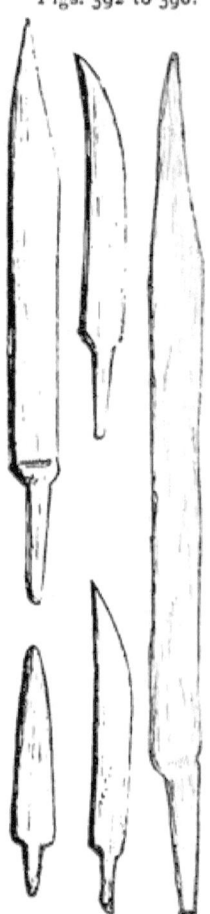

different forms. The most usual shapes are given on figs. 392 to 396. The larger were used for war purposes, the smaller for domestic purposes—the Saxon carrying his own knife with him for his food, attached to his belt, both at home and to the banquets of his friends. The *seax*, as a weapon, is frequently alluded to in *Beowulf*: thus, when Beowulf and the Mother of Grendal, the fiend, were struggling together:—

> " She beset then the hall-guest,
> and drew her *seax*,
> broad, brown-edged."

And in another part, when Beowulf was fighting with the dragon, after having broken his sword in the contest, he

> " Drew his deadly *seax*,
> bitter and battle-sharp,
> that he on his birnie* bore."

Spear and javelin heads are of frequent occurrence; they are of iron, and, although varying considerably, both in size and shape, they all bear a strong and marked resemblance to each other, and have sockets. Their "peculiarity is a longitudinal slit in the socket which received the wooden handle or staff, and which, after being fixed, was closed with iron rings, string-braided, and rivets."† Examples are given in figs. 333 to 403, and again on fig. 404. In interments the spear usually lies by the right side of the skeleton, where the position of the shaft may be traced by a line of decayed wood; at the bottom a metal ferule or ring is sometimes found. The axe is usually of the form here shown, and is of iron. It will be seen how closely some of these resemble the forms found depicted by Anglo-Saxon artists in the MSS. of the period, a selection from which is here given.

The shield appears to have been made of wood, and to have been circular in form. It was frequently covered with leather, and sometimes with thin sheets of bronze. The

* Coat of mail. † C. R. Smith.

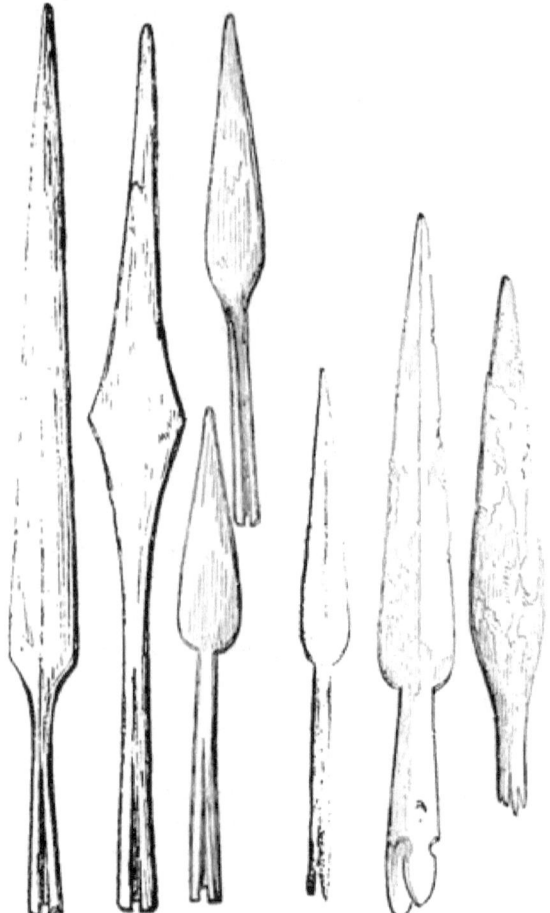

Figs. 397 to 403.

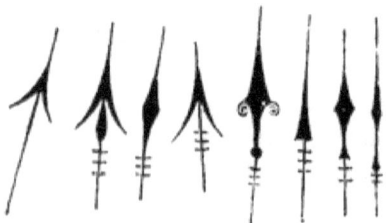

Fig. 404.

boss or umbone was of various forms and sizes, as will be shortly shown. The wood of which the shield was composed appears from Beowulf to have been that of the linden tree:—

> "He seized his shield,
> the yellow linden-wood."

The shield was often called a "war-board;" and we learn that Beowulf, when he was preparing to encounter the fire-dragon, knowing that a wooden shield would be no proof against fire, ordered one "all of iron" to be made for him:—

> "Then commanded he to be made for him
> the refuge of warriors,
> *all of iron*,
> the lord of eorls,
> a wondrous *war-board;*
> he knew right well
> that him forest wood
> might not help,
> linden-wood against fire."

One of the most remarkable remains of shields which has been brought to light is the one at Grimsthorpe,* where, on the breast of the skeleton, lay a mass of decayed wood, a quantity of ferruginous dust—probably the remains of the handle and inside fittings of the shield—and remains of decomposed leather. On these lay two thin plates of bronze, and the umbone or boss of the same metal, which had formed the outer covering of the "war-board." These two plates and the umbone are engraved on fig. 405. The discs or plates of bronze are little thicker than ordinary writing-paper. They each measure twelve and a half inches from point to point, and are three and three quarter inches in width in the middle. They have a raised border of curious design around their outer edge, and they have been,

* See the "Reliquary Quarterly Archæological Journal and Review," vol. ix. p. 180.

like the boss, attached to the shield by pins or rivets. The boss is of very unusual form, and has been attached to the shield by rivets or pins; it is ornamented with engraved lines. From this curious discovery it would appear that this warrior of the Yorkshire Wolds bore a

Fig. 405.

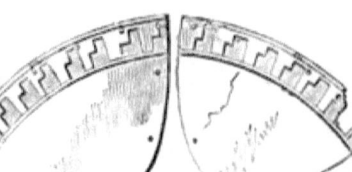

shield formed of wood and covered with leather; that it was faced with plates of bronze, and had a bronze umbone; and that the handle, and probably the strengthening bars, on the inner side were of iron.

Many handles of iron, belonging to shields, have been found in the Kentish and other barrows. The shield, in interments, was usually placed flat on the centre of the body, as shown on fig. 325.

The umbone or boss of the shield was, as I have said,

of various forms. The most usual shapes are, perhaps, those here given from Kentish graves (figs. 406 and 407), and fig. 408, from Tissington, where it was found along

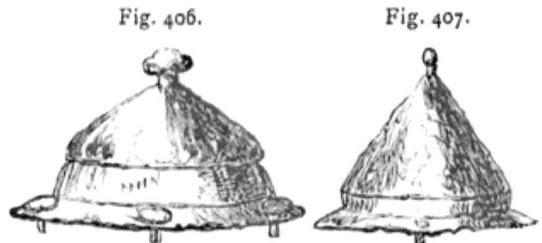

Fig. 406.　　　Fig. 407.

with the sword before described (fig. 372). This extremely interesting relic, which is among the largest ever found, measures nine inches in height. It is, of course, of iron,

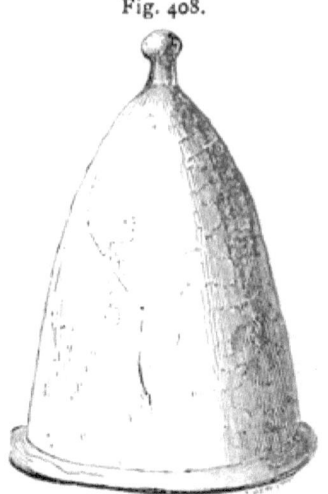

Fig. 408.

and is of the same type as one found at Sibertswold, which is engraved in the "*Inventorum Sepulchrale.*" The texture of cloth in which it had been enfolded when placed by the body of the hero by whom it was borne, is distinctly traceable on several parts of its surface. The umbone, as it lay, was surrounded with the wood, in a complete state of decay, which had once formed the shield; and small fragments of corroded iron, which were doubtless a part of the mountings of the shield, were scattered about.

Of the form of the Anglo-Saxon shield and its umbone, a tolerably good idea may be formed by the series of

Figs. 409 to 416.

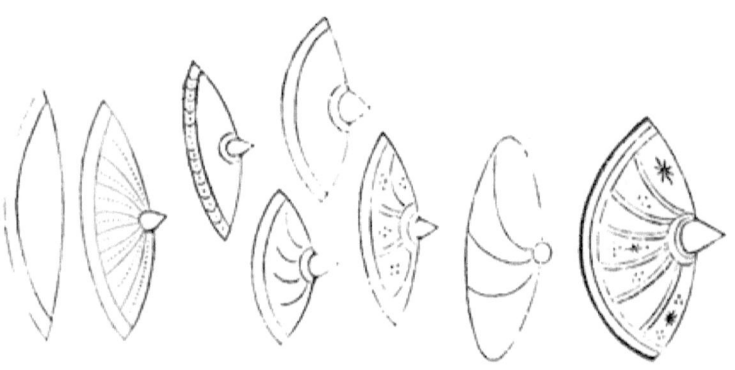

examples here given (figs. 409 to 416), from the illuminated MSS. of the period.

Of Saxon armour the remains yielded to us by the graves are few and far between. Buckles, such as probably fastened the belt or girdle to which the knife, the sword, etc., were suspended, and others which have doubtless belonged to some portions of the dress, are the most abundant. They are of varied form, some being of particularly elegant design, partaking of the character of the fibulæ of the period. Twelve examples from the Kentish graves are given on figs. 417 to 428.

Helmets, or head coverings, in a fragmentary state, have on some few occasions been found. The most remarkable discovery of this kind which has been made is the one

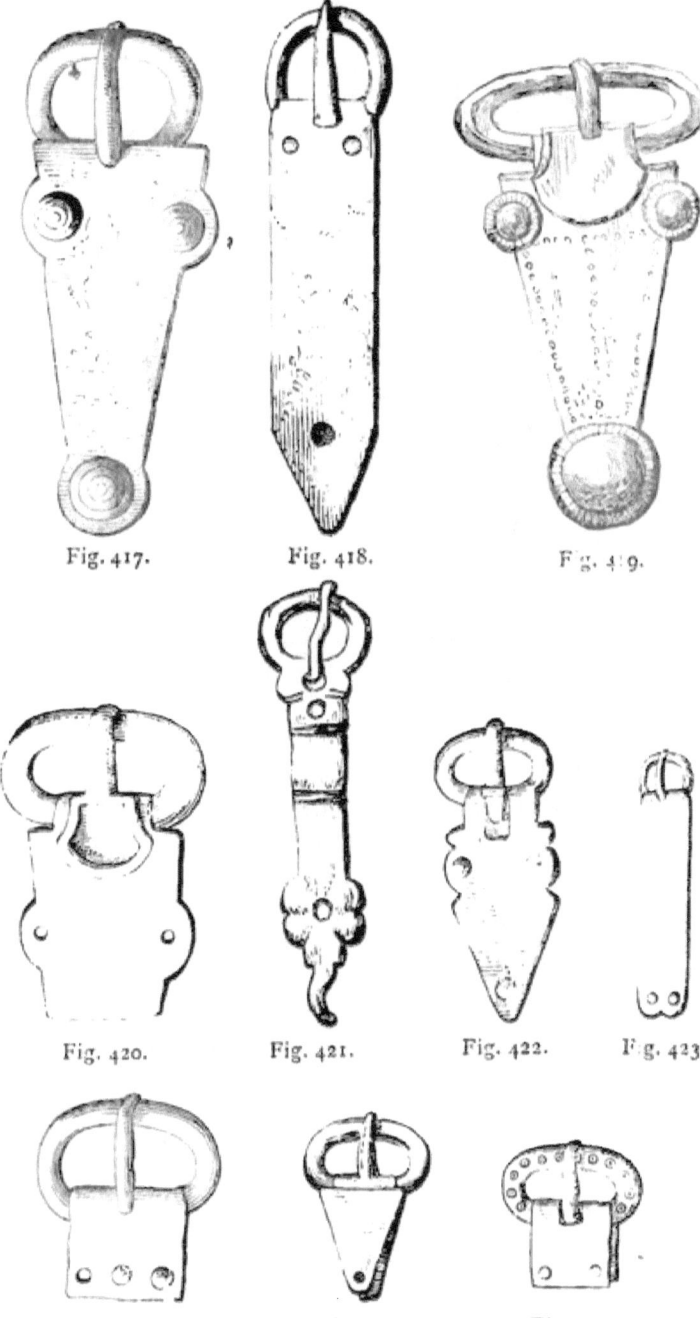

which was found by my friend, the late Mr. Bateman, at Benty Grange,* in Derbyshire, in the year 1848. The

Fig. 427.

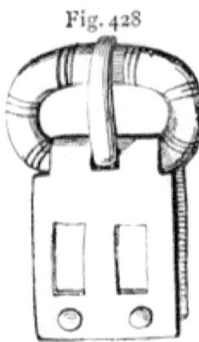

Fig. 428

account of this discovery is so full of interest, and so curious, that I give it in Mr. Bateman's own words. He says :—

"It was our good fortune to open a barrow which afforded a more instructive collection of relics than has ever been discovered in the county, and which are not surpassed in interest by any remains hitherto recovered from any Anglo-Saxon burying-place in the kingdom.

"The barrow, which is on a farm called Benty Grange, a high and bleak situation to the right of the road from Ashbourn to Buxton, near the eighth milestone from the latter place, is of inconsiderable elevation, perhaps not more than two feet at the highest point, but is spread over a pretty large area, and is surrounded by a small fosse or trench. About the centre, and upon the natural soil, had been laid the only body the barrow ever contained, of which not a vestige besides the hair could be distinguished. Near the place which, from the presence of hair, was judged to have been the situation of the head, was a curious assemblage of ornaments, which, from the peculiarly indurated nature of the earth, it was impossible to remove with any degree

* "Ten Years' Diggings," p. 28.

of success. The most remarkable are the silver edging and ornaments of a leathern cup, about three inches in diameter

Fig. 429.

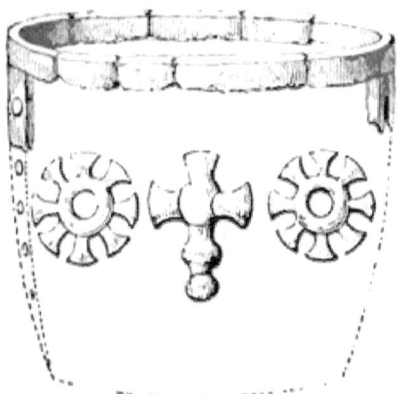

at the mouth, which was decorated by four wheel-shaped ornaments and two crosses of thin silver, affixed by pins of the same metal, clenched inside (fig. 429). The other

Fig. 430.

articles found in the same situation consist of personal ornaments, the chief of which are two circular enamels upon copper 1¾ diameter, in narrow silver frames, and a third, which was so far decomposed as to be irrecoverable (see group, fig. 430); they are enamelled with a yellow interlaced dracontine pattern, intermingled with that peculiar scroll design, visible on the same class of ornaments figured in 'Vestiges,' p. 25, and used in several MSS. of the seventh century, for the purpose of decorating the initial letters. The principle of this design consists of three spiral lines springing from a common centre, and each involution forming an additional centre, for an extension of the pattern, which may be adapted to fill spaces of almost any form. Mr. Westwood has shown in a most able paper in the 40th No. of the Journal of the Archæological Institute, that this style of ornamentation is peculiar to the Anglo-Saxon and Irish artists of the period before stated. The pattern was first cut in the metal, threads of it being left to show the design, by which means cells were formed, in which the enamel was placed before fusion, the whole being then polished became what is known as *champ-levé* enamel. There were also with these enamels a knot of very fine wire, and a quantity of thin bone, variously ornamented with lozenges, etc., which were mostly too much decayed to bear removal; they appeared to have been attached to some garment of silk, as the glossy fibre of such a fabric was very perceptible when they were first uncovered, though it shortly vanished when exposed to the air. Proceeding westward from the head for about six feet, we arrived at a large mass of oxydized iron, which being removed with the utmost care, and having been since repaired where unavoidably broken, now presents a mass of chainwork, and the frame of a helmet. The latter consists of a skeleton formed of iron bands (fig. 431) radiating from the crown of the head, and riveted to a circle of the same metal which encompassed the brow: from the impression on the metal it is evident that the outside was covered with plates of horn disposed diagonally so as to produce a herring-bone pattern; the ends of these plates were secured beneath with strips of horn corresponding

Fig. 431.

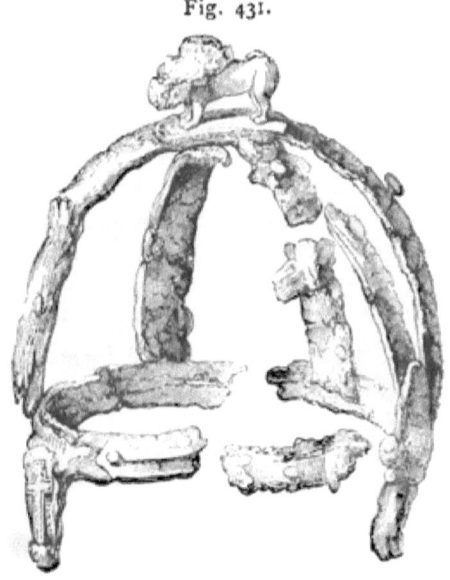

with the iron framework, and attached to it by ornamental rivets of silver at intervals of about an inch and a half

Fig. 432.

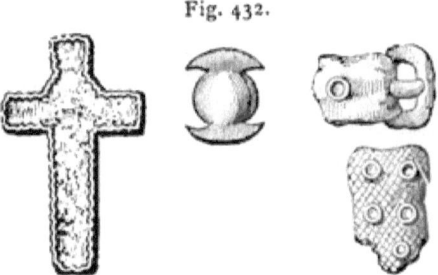

from each other. On the bottom of the front rib, which projects so as to form a nasal, is a small silver cross, slightly ornamented round the edges by a beaded moulding; and on the crown of the helmet is an elliptical bronze plate supporting the figure of an animal carved in iron, with bronze eyes, now much corroded, but perfectly distinct as the representation of a hog. There are, too, many fragments, some more or less ornamented with silver, which have been riveted to some part of the helmet in a manner not to be explained or even understood; there are also some small buckles of iron, which probably served to fasten it upon the head. Amongst the chainwork is a very curious six-pronged instrument of iron, in shape much like an ordinary hay-fork, with the difference of the tang, which in the latter is driven into the shaft, being in this instrument flattened and doubled over so as to form a small loop, apparently convenient for suspension; whether it belonged to the helmet or the corselet, next to be described, is uncertain. The iron chainwork already named consists of a large number of links of two kinds, attached to each other by small rings, half an inch in diameter; one kind are flat and lozenge-shaped, about an inch and a half long; the others are all of one kind, but of different lengths, varying from four to ten inches. They are simply lengths of square rod iron with perforated ends, through which pass the rings connecting them with the diamond-shaped links; they all show the impression of cloth over a considerable part of the surface, and it is, therefore, no improbable conjecture that they would originally constitute a kind of quilted cuirass, by being sewn up within, or upon, a doublet of strong cloth. The peculiarly indurated and corrosive nature of the soil in this barrow is a point of some interest, and it will not be out of place to state that such has generally been the case in tumuli in Derbyshire, where the more important Saxon burials have taken place, whilst the more ancient Celtic interments are generally found in good condition, owing to there having been no special preparation of the earth, which in these cases has undergone a mixing or tempering with some corrosive liquid, the result of which is the presence of thin ochrey veins in the earth, and the decomposition of nearly the whole of the human remains. The following extract from Professor Worsaae's 'Antiquities of Denmark' illustrates the helmet, which is the only example of the kind hitherto discovered, either in this country or on the Continent :—

"'The helmets of the ancient Scandinavians, which were furnished with crests, usually in the form of animals, were probably in most cases only the skins of the heads of animals, drawn over a framework of wood or leather, as the coat of mail was usually of strong quilted linen, or thick woven cloth.'"

To this the translator of the English edition appends the important information, that "the animal generally represented was the boar; and it is to this custom that reference is made in *Beowulf*, where the poet speaks of the boar of gold, the boar hard as iron."

"Spýn eal ȝylben.
Eoꝼeꞃ Iꞃen heꝛð"

Nor are allusions to this custom of wearing the figure of a boar—not in honour of the animal, but of Freya, to whom it was sacred—confined to Beowulf; they are to be found in the *Edda* and in the *Sagas;* while Tacitus, in his work, "De Moribus Germanorum," distinctly refers to the same usage and its religious intention, as propitiating the protection of their goddess in battle. As a further illustration, not only of the helmet, but also of the chainwork, the following extracts from Beowulf are transcribed from Mr. C. R. Smith's "Collectanea Antiqua," vol. ii., p. 240 :—

eoꞃep-lic ꞃcíón oꞃ-oꞃeꞃ hleoꞃ bæꞃon; ᵹe-hꞃoben ᵹolde, ꞃáh anꝺ ꞃýꞃ-heaꞃꝺ, ꞃeꞃh peaꞃꝺe heólꝺ.	"They seemed a boar's form to bear over their cheeks; twisted with gold, variegated and hardened in the fire, this kept the guard of life: l. 604.
be-ꞃonᵹen ꞃꞃeá-ppáꞃnum, ꞃpa híne ꞃýpn-haᵹum poꞃhꞇe pæpna ꞃmíð, punꝺꞃum ꞇeóde, be-ꞃeꞇꞇe ꞃpín-lícum, þ híne ꞃýðan nó bꞃonꝺ né beaꝺo-mecaꞃ bíꞇan ne meahꞇon:	Surrounded with lordly chains, even as in days of yore the weapon smith had wrought it, had wondrously furnished it, [swine, had set it round with the shapes of that never afterwards brand or war-knife might have power to bite it: l. 2901
æꞇ þæm áde pæꞃ eþ-ᵹe-ꞃýne ꞃpáꞇ-ꞃah ꞃýꞃce, ꞃpýn eal-ᵹýlden, eoꞃeꞃ íꞃen heapꝺ:	At the pile was easy to be seen the mail shirt covered with gore, the hog of gold, the boar hard as iron: l. 2213.
Dáꞇ ðá m-beꞃan eoꞃoꞃ-heáꞃoð-ꞃeᵹn, heaðo-ꞃꞇeapne helm, heꞃe-býꞃnan, ᵹúð-ꞃpeoꞃꝺ ᵹeáꞇo-lic:	Then commanded he to bring in the boar, an ornament to the head, the helmet lofty in war, the grey mail coat, the ready battle sword." l. 4299.

It will be noticed in these extracts that "mail coat" or "mail shirt" is twice mentioned, as well as the "helmet lofty in war." Thus the passages in a remarkable degree illustrate this extraordinary discovery, which embraced a coat of mail along with the helmet and other objects. The coat appears to have consisted of a mass of chainwork, the links of which were attached to each other by small rings.

Fragments of another helmet were the following year found in another barrow in the same neighbourhood, at Newhaven, along with other objects of interest. The barrow had, however, at some previous time been grievously mutilated. Of this barrow Mr. Bateman says: "We opened a mutilated mound of earth in a field near Newhaven House, called the Low, two-thirds of which had been removed, and the remainder more or less disturbed, so that nothing was found in its original state, which is much to be regretted, as the contents appear to be late in date, and different in character from anything we have before found in tumuli. The mound itself, being constructed of tempered earth, bore some analogy to the gravehill of the Saxon Thegn, opened at Benty Grange about a year before, and, like it, was without human remains, if we except a few fragments of calcined bone, which are too minute to be certainly assigned either to a human or animal subject. The articles found comprise many small pieces of thin iron straps or bands, more or less overlaid with bronze, which are by no means unlike the framework of the helmet found at Benty Grange. There is also a boss of thin bronze, three inches diameter, pierced with three holes for attachment to the dress, (?) and divided by raised concentric circles, between which the metal is ornamented with a dotted chevron pattern, in the angles of which are small roses punched by a die. Another object in bronze is a small round vessel or box of thick cast metal, surrounded by six vertical ribs, and having two perforated ears, serving probably better to secure the lid and suspend the box. Although it measures less than an inch in height, and less than two in diameter, it weighs full 3½ ounces. A similar box, with the lid, on which is a cross formed of annulets, found with Roman remains at Lincoln, is engraved at page 30 of the Lincoln Book of the Archæological Institute, where it is called a pyx. Two others, discovered at Lewes, are engraved in the 'Archæologia,' vol. xxxi., page 437, one of which has the lid bearing a cross precisely similar to the Lincoln example, whence it is certain that they must be assigned to a Christian period, probably not long previous to the extinction of the Saxon monarchy. The last object there is occasion to describe is an iron ferrule or hoop, an inch and a half in diameter, one edge of which is turned inwards, so as to prevent its slipping up the shaft on which it has been fixed. We also found some shapeless pieces of melted glass, which, from their variegated appearance, might be the product of fused beads; and observed many pieces of charred wood throughout the mound, which may possibly not have been of a sepulchral character."

Another helmet, or defensive cap, was found some years ago at Leckhampton Hill, in Gloucestershire, the ribs of which bear a striking analogy to the one here described.

A remarkable discovery, which included portions of what is very plausibly considered to be a helmet, was also, a few years ago, made on the estate of Mr. Francis Wedgwood, at Barlaston, in Staffordshire. The particulars of this I now for the first time make public. The grave, which was seven feet in length by two feet in width, was cut in the solid red-sandstone rock. It was about fifteen inches in depth at the deepest part, which was at the south-east corner, and died out with the slope of the hill towards the north-west, and the earth which covered it (which had probably been tempered in the usual manner) was only a few inches in thickness. It was on the slope of the hill. At the upper or northerly end of the grave a basin-like cavity, two or three inches in depth, was cut in the floor of rock (see A in the plan, fig. 433). In this hollow, which had evidently been intended for the helmeted head of the deceased to rest in, was found the remains of what I have alluded to as justly considered to be remains of a bronze helmet. The skeleton had, as is so frequently the case in Anglo-Saxon interments, entirely disappeared, but on its right side lay the sword (B), and on the left a knife (C).

The fragments in the cavity consisted of several pieces of curved bronze, highly ornamented, which had probably, with other plain curved pieces, formed the framework of the helmet; some thin plates of bronze; a flat ring of bronze, beautifully ornamented (fig. 434), which is conjectured to have been the top of the framework of the helmet; and three enamelled discs, of a similar character to what have been

Fig. 433.

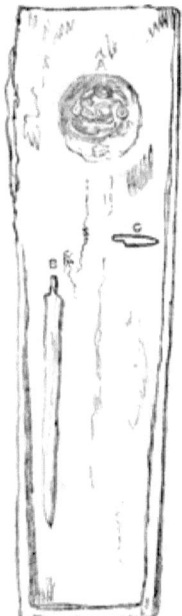

Fig. 434.

elsewhere found, with hooks for suspension, or attachment to leather or other substance. One of these is engraved, of its real size, on the next illustration (fig. 435). The centre is of enamel mosaic work, ground down level with the metal, as in the old Chinese enamels.

Fig. 435.

The inference to be drawn from this curious discovery is, that the grave was that of a Saxon of high rank, who had been buried in his full dress, and that the cavity had been specially cut out in the floor of the rock grave to admit of the helmet being worn as when he was living. No remains of a shield were noticed, nor were any other remains found in the locality, which was carefully dug over for the purpose.

Enamelled discs, or pendants, such as I have just spoken

of (see fig. 435), have been occasionally found in other
localities, as will have been noticed in the course of the
last few pages. The use of these curious objects is very
obscure, and I am not aware that any very particular
attention has been paid to them. Portions of these were
found in the Benty Grange barrow (fig. 430), along with the
Saxon helmet. A very perfect example was found in a barrow on Middleton Moor, Derbyshire, in 1788,* where it was

found lying near the shoulder. In the same barrow was a
portion of another enamelled ornament, the iron umbone
of a shield, and a thin vessel of bronze—described as like
a shallow basin—which probably formed a portion of a hel-

* "Vestiges," p. 24.

met. These two interesting relics are here engraved (figs. 436 and 437). The first of these will be seen to bear a striking resemblance to the Barlaston example (fig. 435), and the second, in form, to be very similar to the next example (fig. 438), from the museum of the Royal Irish Academy. Some precisely similar objects—similar in design and in size to figs. 435 and 436—were found at Ches-

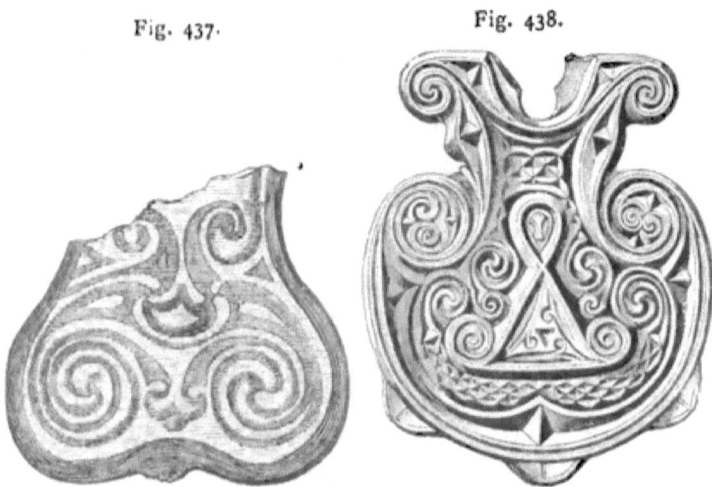

Fig. 437.   Fig. 438.

terton. Of the four of these objects there found, two are precisely alike, and had hooks for suspension in the same manner as is shown on fig. 435; the other two have no hooks, and are of a different pattern of enamelling. Other examples have been brought to light in different localities, but these will be sufficient for my present purpose.

It is, of course, very difficult to come to any conclusion, in the present state of our knowledge of Anglo-Saxon history, as to the original uses of these and other objects. That these enamelled and handled discs were intended for suspension by their hooks there can be but little doubt, and it seems not improbable that they might serve as pendants to the helmet; the two with hooks possibly hanging as ear-guards or coverings, and the others being attached by pins or rivets to, perhaps, the front and back of the circle. It is hoped that ultimately the use of these curious relics may be correctly ascertained. In the barrow at Grimthorpe, already referred to, a disc of somewhat similar character,

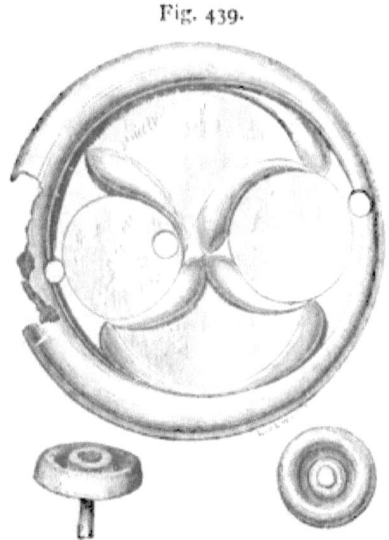

Fig. 439.

of thin metal, was found. It had been attached by three pins or rivets, the holes for which remained. It was not enamelled, but decorated with raised ornaments. It is engraved of its full size on fig. 439.

A singular plate of cast and chased bronze, strongly gilt, and set with garnets, found in Northamptonshire, and now in the Bateman museum, is engraved in the "Reliquary," vol. i. It has at the back, besides a central projection, four pierced projections for attachment to leather or other substance, and four "swivel" projections, if they may be so termed, on its edges, to which other matters have been attached by rivets, which are still remaining.

Enamelling and goldsmiths' work were evidently arts in which the Anglo-Saxon artificers excelled; some of the rings and fibulæ, and other relics, being of extreme elegance and richness, and of great beauty in design.

Having spoken of the arms, helmets, etc., found in Anglo-Saxon graves, it will be well before proceeding to describe the personal ornaments, to note that horse-shoes are occasionally met with in interments, showing that the horse was, in some instances, buried with its rider. Having given, on fig. 324, the form of a horse-shoe of the Romano-British period, I now engrave examples of those of the Anglo-Saxon times. Figs. 440 and 441 are two shoes from a Saxon grave in Berkshire.

Fig. 440.   Fig. 441.

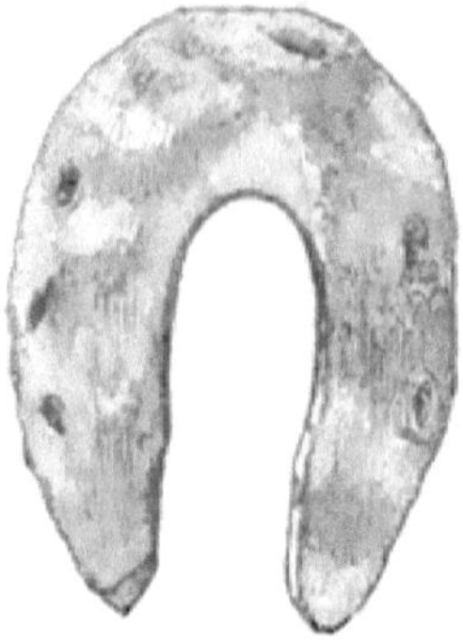

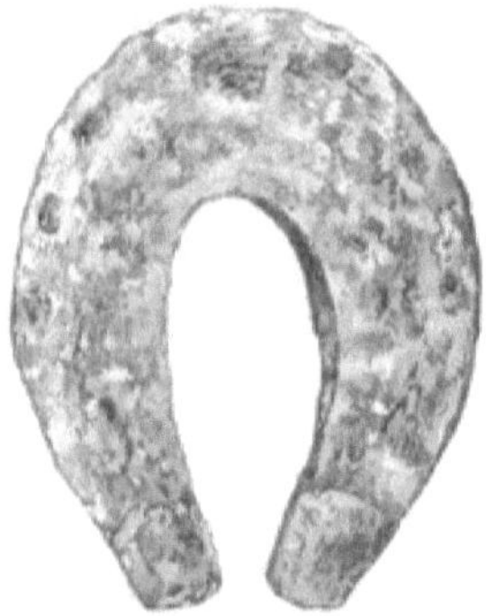

They will be seen to be of a very different form to those of the preceding era. One has calkins, but the other is without, and both are even on the outer edge, not "bulged," as those of Roman times are. In illustration of this matter, I am enabled, through the courtesy of my friend Mr. Fleming, to give the accompanying engraving from his admirable work on "Horse-shoes and Horse-shoeing," to which I would direct the attention of all who are interested in this branch of archæological inquiry. The engraving represents the contents of a grave-mound excavated at Selzen, on the Rhine, by Lindenschmidt, in which, along with the skeleton of the warrior, were the skull and other remains of his horse, with portions of horse-shoes, as well as some urns of good character, and of close resemblance to those of our Anglo-Saxon period. Tumuli containing the remains of horses are of unfrequent occurrence in England, and therefore this example becomes interesting as an illustration for comparison.

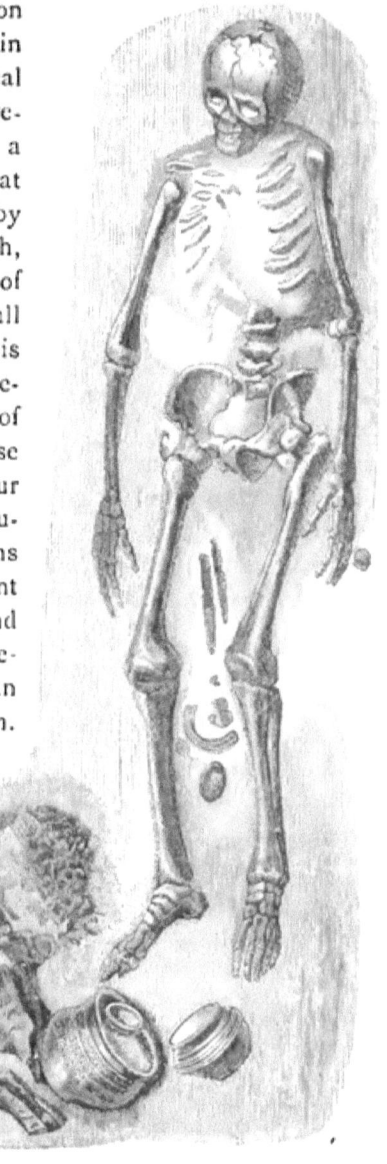

CHAPTER XV.

Anglo-Saxon Period—Fibulæ—Enamelled circular Fibulæ—Gold Fibulæ—Pendant Cross—Cruciform Fibulæ—Penannular Fibulæ—Irish and English examples—Pendant Ornaments, etc.

THE fibulæ of the Anglo-Saxon period are the most remarkable, perhaps, of any of the products of the grave-mounds of that people. They are of extreme interest, not merely from their design and the excellence of their workmanship, or from their various forms and styles of ornamentation, but because by their varieties the different races to which they belonged can, in great measure, be determined.

The more beautiful and elaborate, and at the same time the richest in effect, of these various forms of fibulæ are those of circular form, which, although found in various parts of the kingdom, are more abundant in the barrows of Kent than elsewhere. The finest of these ever discovered was found in 1771, " near the neck, or rather more towards the right shoulder," of the female skeleton in a grave six feet deep, ten feet long, and eight feet wide, on Kingston Down, along with some small silver fibulæ, a golden amulet, some small hinges, a chain, some bronze vessels, pottery, and a variety of other articles. This fibula, here engraved (fig. 443), which is quite unique, "stands at the head of a class by no means extensive, characterized by being formed of separate plates of metal, enclosed by a band round the edges. The shell of this extraordinary brooch is entirely of gold. The upper surface is divided into no less than seven compartments, subdivided into cells of various forms. Those of the first and fifth are semi-circles, with a peculiar graduated figure, somewhat resembling the steps or base of a cross, which also occurs in all the compartments, and in four circles, placed crosswise with triangles. The cells within this step-like figure

Fig. 443.

and the triangular are filled with turquoises; the remaining cells of the various compartments with garnets, laid upon gold-foil, except the sixth, which forms an umbo, and bosses in the circle, which are composed apparently of mother-of-pearl. The second and fourth compartments contain vermicular gold chain-work, neatly milled and attached to the ground of the plate. The reverse of the fibula is also richly decorated."

The vertical hinge of the acus is ornamented with a cross set with stones, and with filigree work round its base. The clasp which receives the point of the acus is formed to represent a serpent's head, the eyes and nostrils of which, and the bending of the neck, are marked in filigree. This precious jewel was secured by a loop which admitted of its being sewn upon the dress.

Another remarkably fine example, found on the breast of a female skeleton in Berkshire, is now in the Ashmolean museum. It measures two inches and seven-eighths in diameter. The base is formed of a thin plate of silver, above which, resting apparently on a bed of paste, is a plate of copper, to which is affixed a frame-work of the same metal, giving the outline of the pattern. The four divisions of the exterior circle were originally filled with paste, on which were laid thin laminæ of gold, ornamented with an interlaced pattern in gold wire of two sizes, delicately milled or notched, resembling rope-work. Of these compartments one is now vacant. This wire ornament was pressed into the gold plate beneath, and there are no traces of any other means than pressure having been used to fix it. The four smaller circles and that in the centre are ornamented with bosses of a white substance, either ivory or bone, but the material is so much decomposed it is difficult to say which; these bosses are attached to the copper plate beneath by iron pins. The entire face of the fibula was originally set with small pieces of garnet-coloured glass laid upon hatched gold-foil. The upper and lower plates of this ornament are bound together by a band of copper gilt, slightly grooved. The acus is lost.

The magnificent circular fibula of gold here engraved (fig. 444) was discovered some years ago in a barrow on Winster Moor, in Derbyshire. It was formed of gold filigree work, which was mounted on a silver plate. It was set with stones or paste on chequered gold-foil, and measured two

inches in diameter. Along with this fibula were found the following interesting articles: a cross of pure gold, orna-

mented, like the fibula, with filigree work, and having a garnet cut in facets set in its centre (fig. 445); a silver armlet; two glass vessels, and a number of beads. These and some other articles were all found by the sides of two cinerary urns.

Many of the circular fibulæ are, of course, of a much smaller and less elaborate character than those here given.

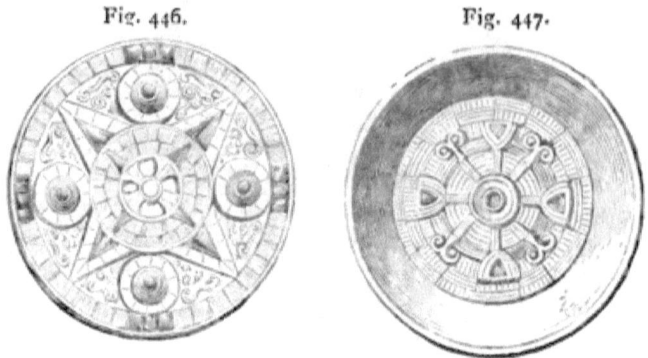

Fig. 446.   Fig. 447.

They all, however, bear, exclusive of the fact of their being found along with other evidences of the period to which

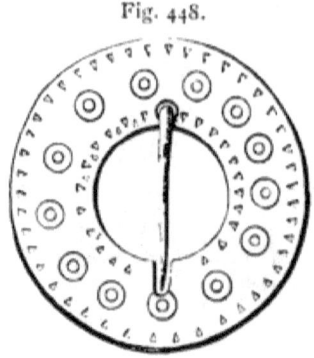

Fig. 448.

they belong, characteristics which cannot well be mistaken.

These circular fibulæ appear to have been worn by the

Anglo-Saxon ladies on the breast or, occasionally, shoulder. They were probably, therefore, used for fastening the dress on the bosom, as is so often seen in illuminated MSS. and on tombs of a later period.

Another extensive class of Anglo-Saxon fibulæ are what are usually called, though not very satisfactorily, cruciform, or cross-shaped. Fibulæ of this class are, perhaps, most abundant in the midland and south-eastern counties, but they are of very rare occurrence in Kent. They would appear, therefore, to have appertained mostly to the Angles, who were the inhabitants of Mercia, East Anglia, and Northumbria. They are sometimes of silver, but usually of bronze, and are variously ornamented with interlaced

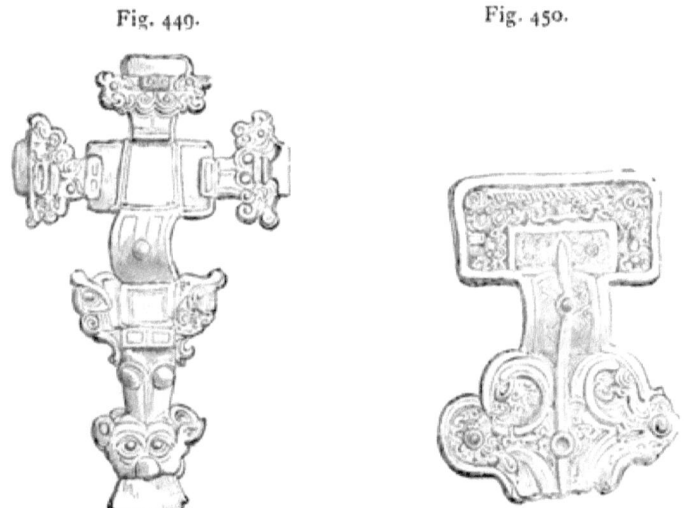

Fig. 449.   Fig. 450.

work, heads, and borders of various designs. Their form will be best understood from the accompanying engravings, which exhibit some of the most usual varieties. They are from Northamptonshire, Leicestershire, Suffolk, and Cambridgeshire, and will serve as typical examples of this class of brooch.

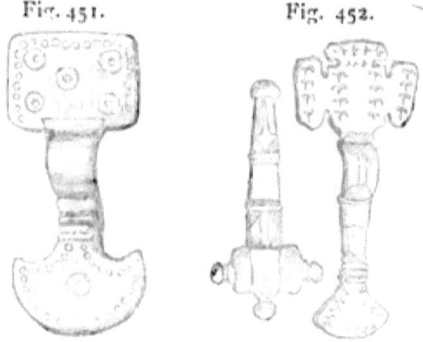

Fig. 451.   Fig. 452.

Another totally distinct kind of fibula, or brooch, which is considered to be peculiarly of Irish type, but which, nevertheless, is occasionally met with in England, remains to be noticed. I allude, of course, to brooches of the penannular

Fig. 453.

form,* the general type of which will be understood by the engravings given on figs. 453, 454, and 455, which are all Irish examples of more or less decorative character. The originals are in the museum of the Royal Irish Academy,

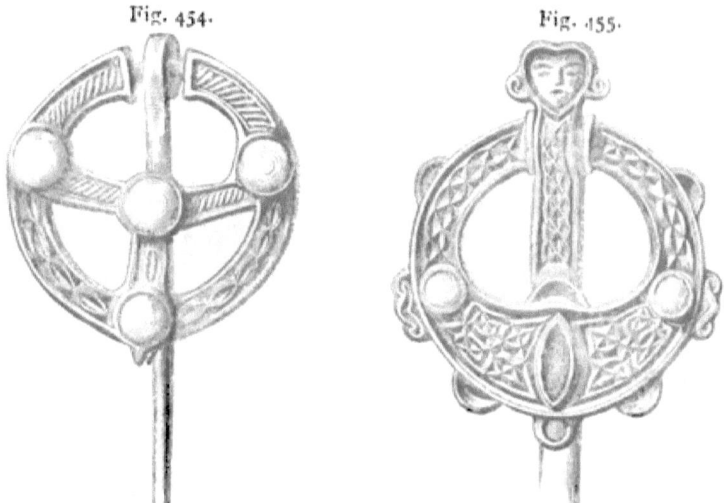

Fig. 454.   Fig. 455.

as are also many other exquisite specimens of these interesting examples of early art.

The one next figured (fig. 456) was discovered in Westmoreland, and described and engraved in the *Archæological Journal*, vol. ix. page 90. This beautiful fibula I here engrave of a reduced size. The ring, it will be seen, moves freely round the upper half of the brooch, the lower or flat part of which is divided so as to allow of the passage of the acus through it. "It is set with flat bosses, five on either side. Each of these flat dilated parts of this curious ornament appears to proceed from the jaws of a monstrous head, imperfectly simulating that

* For a more extended and fully illustrated account of penannular brooches, the reader is referred to the "Reliquary," vol. iii.

of a serpent or dragon; and between the jaw is introduced the intertwined triplet, or *triquetra*, the same ornament which is found on the sculptured cross at Kirk Michael,

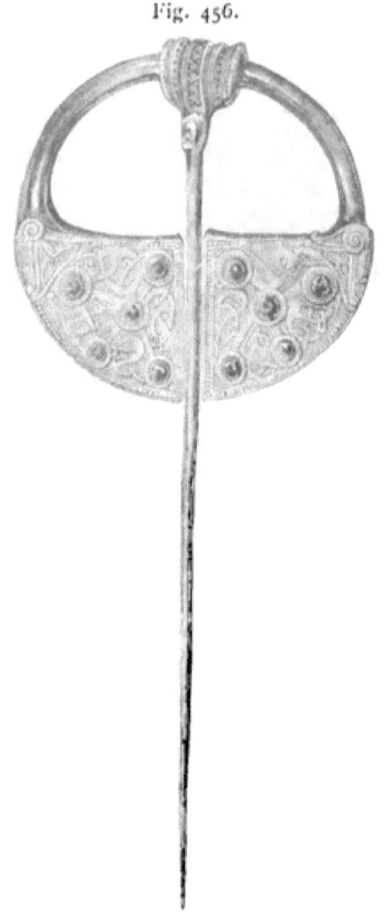

Fig. 456.

Isle of Man, and on some Saxon coins." This example is of silver. With it was found a silver armlet—a simple twisted bar of decreasing thickness towards the extremities,

Fig. 457.

which are hooked. The dimensions of the fibula are, length of acus, eleven inches; greatest diameter of circular part, five inches; width of the dilated part, two inches; weight, 8 oz. 8 dwt.

By far the finest example found in England is the one next figured (fig. 457). It was found in 1862, near the picturesque village of Bonsall, in the High Peak of Derbyshire. It is of bronze, and is here engraved of its full size. The ring measures three inches and seven-eighths in its greatest diameter, and the acus, which is not engraved of its full length, is six inches and three-quarters long.

It has originally been set with amber or paste, and has been richly gilt and enamelled. The interlaced ornaments are most exquisitely and elaborately formed, and are of great variety, and the heads of animals are of excellent and characteristic form. The head of the acus, or pin, is large and beautifully ornamented, and, like the ring, has been set with studs. The pin itself, as will be seen by the accompanying engraving (fig. 458), is flattened and made thin at its upper end, and bent so as to allow of the free passage of the ring through it, and is riveted on to the ornamented plate in front.

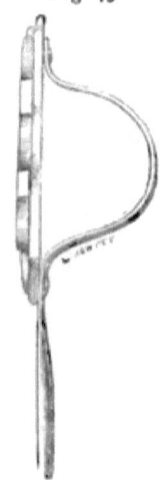

Fig. 458.

It is remarkable that, in this fibula, the ring, which, like other examples of this form of brooch, has been made to play freely for half its circumference through the acus, has been riveted to the head of the pin in the position shown in the engraving. That it has been much worn in this position—*across* the breast or shoulder—is evident from the ring being much worn where the pin has pressed against it when clasped. I believe this is the only example on record in which the pin has been *fixed* to the side of the ring, and this was certainly not the original intention of the maker of the brooch, but was done subsequently. This will be seen by the engraving of the profile of the head of the acus, on fig. 458. On one or two examples of penannular brooches, inscriptions in Ogham

Fig. 459.

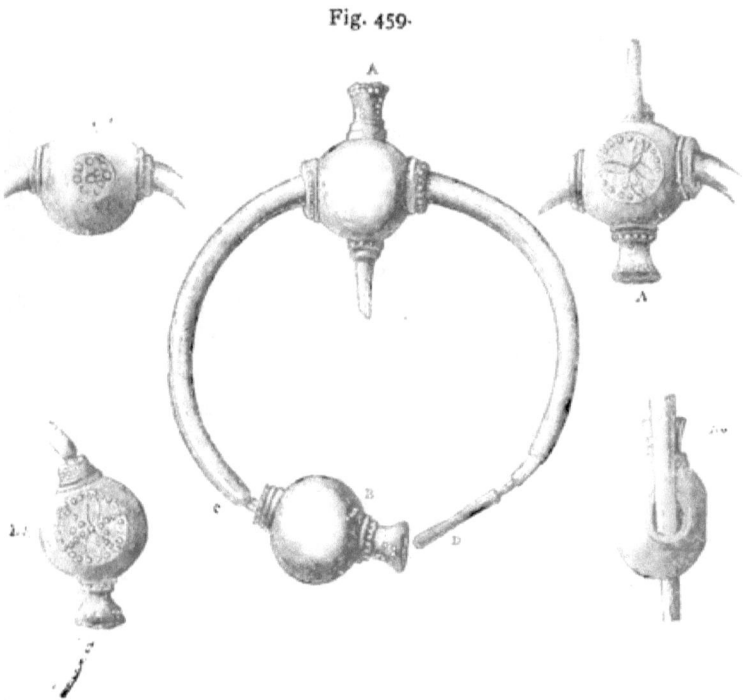

characters have been found, and it is highly interesting to be able to add that, on the back of the Derbyshire example, faint traces of Oghams still remain.

Another brooch, of silver, found in England, though different in form from the expanded examples just given, and although of later date, is nevertheless of the same construction. It is engraved of a reduced size on fig. 459. "The acus has been broken off. There appears to have been a third knob, now lost, which should correspond with the knob B, the acus passing between the two. The upper knob A is very loose, and moves freely around the ring. The knob B turns, but much less freely, and does not pass over C, having merely a lateral motion of one-fourth of an inch." The diameter of the widest part is nearly five and a half inches ; the globular ornaments measure one and a quarter inches in diameter. The under side of each of the balls is flat, and is engraved with ornaments, as shown on the engraving. This brooch belongs to Mr. C. Carus Wilson, and closely resembles some of the Irish examples.

Of the mode of wearing penannular brooches, the late Mr. Fairholt says: " By the sumptuary laws of the ancient Irish, the size of these brooches, or fibulæ, were regulated according to the rank of the wearer. The highest price of a *silver* bodkin for a king or an *ollamh*, which, according to Vallancy, was *thirty* heifers, when made of refined silver; the lowest value attached to them being the worth of three heifers. From this it may be inferred, that the rank of the wearer might always be guessed at from the fibulæ he wore." The rank of the wearers of the "Tara Brooch"—the most famous of all the Irish brooches at present known—and of the Derbyshire example, must, judging from their large size and truly exquisite workmanship, have been high.

The extreme rarity of brooches of this form in England, leads one, naturally, to the conclusion that they were not much worn by the inhabitants of this country, and that, therefore, they can hardly be considered to belong to the nationality, if I may so speak, of the Anglo-Saxons. Nevertheless, examples having been here found in close proximity to undoubted Anglo-Saxon remains, and the style of ornamentation being strictly in keeping with much belonging to that period, there can be no doubt that they must be included amongst our Anglo-Saxon antiquities.

Some of the most beautiful objects, along with the fibulæ,

which the graves of the Anglo-Saxons yield, are the pendant ornaments of various kinds which were worn by that race of people. The objects of this class are extremely varied; but their beauty, like those of the richly studded and gilt fibulæ, and the enamelled studs and bosses, cannot well be understood without the aid of coloured illustrations. Of these a set of exquisite pendants were found along with several other interesting objects, in a barrow on Brassington Moor, by Mr. Bateman. Eleven of these pendants are large and brilliantly coloured garnets beautifully set in pure gold, two are entirely of gold, and the third, also of gold, is of spiral wire. Two beads, one of green glass, the other of white and blue glass, were also found.

Gold drops of a similar character to those just described have been frequently found in the Kentish graves, as have also one or two crosses very similar to the one engraved on a previous page (fig. 445). Circular pendants of gold and other materials, decorated with enamelled or raised interlaced and other ornaments, or set with garnets and other stones, are also found. Among the most interesting of this class of pendant ornaments are coins to which loops have been attached. Examples have been found in Kent and elsewhere, and show that the fashion to some extent indulged in at the present day of wearing coins attached to watch chains, etc., is at least of Anglo-Saxon origin.

CHAPTER XVI.

Anglo-Saxon Period—Buckets—Drinking-cups of wood—Bronze Bowls—Bronze Boxes—Combs—Tweezers—Châtelaines—Girdle Ornaments—Keys—Hair-pins—Counters, or Draughtmen, and Dice—Querns—Triturating Stones, etc.—Conclusion.

BUCKETS, so called, and very appropriately, from their close resemblance in form to our modern vessels bearing that name, are occasionally found in Anglo-Saxon graves. They are small wooden vessels bound round with hoops or rims of bronze, more or less ornamented, and have a handle of the same metal arched over their tops. Of course in every case the wooden staves of which they were composed, and which were of ash, are decomposed, the hoops, handle, and mountings alone remaining. They vary very much in size; one from Bourne Park had the lower hoop twelve inches in diameter, and the upper one ten inches, and the whole height appears to have been about a foot; the handle was hooked at its ends exactly the same as in our present buckets, and fitted into loops on the sides; it had three looped bronze feet to stand upon. Other examples only measure four or five inches in diameter. The example here engraved (fig. 460) was found in Northamptonshire, along with other remains. It is composed of three encircling hoops of bronze, and has its handle and attachments also of the same metal.

The next example (fig. 461) is from Fairford, in Gloucestershire, and is three inches in height, and four inches in diameter. The hoops and mountings are of bronze.

Fig. 460.

Fig. 461.

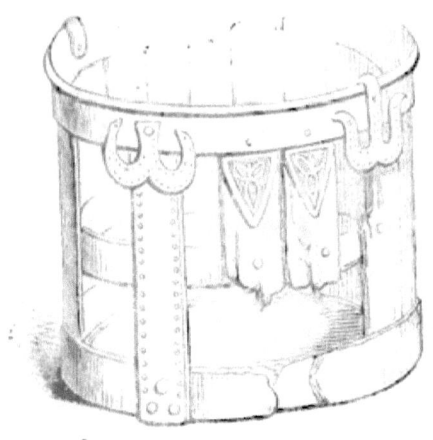

Another example, which I give for the purpose of comparison, is from Envermeu, in Normandy (fig. 462). Of

Fig. 462.

the use of these utensils nothing certain, of course, is known, but it is conjectured they were used for bringing in mead, ale, or wine, to fill the drinking-cups—the objection to this as a general rule being their very small size. "The Anglo-Saxon translation of the Book of Judges (vii. 20) rendered *hydrias confregissent*, by 'to-bρǽcon þa bucaρ,' *i.e.* 'they broke the buckets.' A common name for this vessel, which was properly called *buc*, was *œscen*, signifying literally a vessel made of ash, the favourite wood of the Anglo-Saxons."

Drinking-cups were sometimes of wood. Of these, two examples are here given. The first of these has a rim of brass, the second a like rim attached by overlapping bands.

DRINKING-CUPS AND BOWLS. 283

It has also a number of small bands of the same metal riveted on to mend cracks in the wood. They were found in a barrow on Sibertswold Down, in Kent.

Fig. 463.

Bowls of bronze are occasionally also found. Some of these are plain, others enamelled or otherwise ornamented,

Fig. 464.

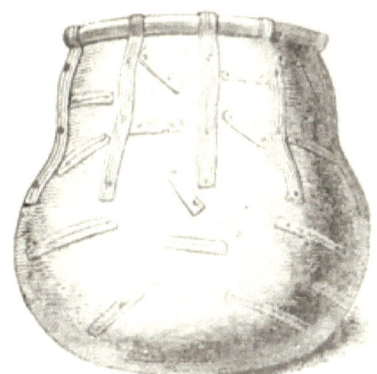

and others, again, gilt. Many of them appear from their form to have been of Roman origin. Some remarkably fine examples have been yielded by the graves of Kent and other districts. The one here engraved (fig. 465) was found at Over-Haddon, in Derbyshire, along with the remains of a circular enamelled disc of the kind described

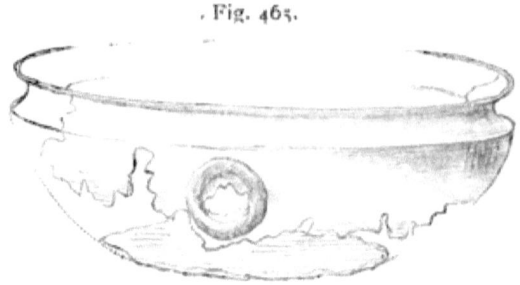

Fig. 465.

on a previous page, and other relics. The bowl was seven inches in diameter, and had originally two handles. They are supposed to have been used for placing hot meats in, on the table. They range in size from four or five to twelve or fourteen inches in diameter.

Small boxes of bronze are occasionally found, and are of different forms. Some are plain upright boxes with lids, just intended to hold sewing materials—in fact, the workboxes of the Saxon ladies—and others are rather large, and have been intended to contain the comb, etc.: they are, therefore, a kind of dressing-cases. The box engraved on fig. 466 was found along with other Saxon remains near Church Sterndale. The grave, which was cut in the rock, contained a skeleton of a woman; the lower bones were fairly preserved, but of the upper parts there were but few remains, the enamel crowns of the teeth being in the best condition. " At the left hip was a small iron knife four inches long, and where the right shoulder had been was an assemblage of curious articles, the most important of which was a small bronze box or canister, with a lid to slide on, measuring altogether two inches high, and the same in diameter. When found, it was much crushed, but still retained, inside, remains of thread, and bore on the outside impressions of linen cloth. Close to it were two bronze pins or broken needles, and a mass of corroded iron, some

Fig. 466.

of which has been wire chainwork connected with a small bronze ornament with five perforations, plated with silver, and engraved with a cable pattern, near which were two

Fig. 467.

iron implements of larger size, the whole comprising the girdle and châtelaine, with appendages, of a Saxon lady. Many pieces of hazel stick were found in contact with these relics, which were probably the remains of a basket in which they were placed at the funeral. All the iron shows impressions of woven fabrics, three varieties being distinguishable; namely, coarse and fine linen, and coarse flannel or woollen cloth. The box is very faintly ornamented by lozenges, produced by the intersection of oblique lines scratched in the metal."

The next engraving shows a bronze box of quite a different character, found with Anglo-Saxon remains at New-

haven. It is two inches in diameter, but very thick. It has six vertical ribs and two bars for attachment of the lid.

Needles and pins are frequently met with. The two shown on fig. 466 will, however, be sufficient to call attention to these minute objects.

Combs of the Anglo-Saxon period differ but little from those of the Romans, or indeed from those of the present day. They were, both Roman and Saxon, sometimes toothed on one side and sometimes on both sides, and were made alike of wood, of metal, of bone, and of ivory. Boxwood appears to have been so much used for the manufacture of combs as to have occasionally given its own name to them. Thus Martial says:—

> "Quid faciet nullos hic inventura capillos,
> Multifido buxus quæ tibi dente datur?"

Wooden combs have naturally for the most part perished, but fragments have occasionally been found. Combs, both of bronze and iron, of the Roman period, have also been discovered. The greater part, however, both of that and of the Saxon period, which have been exhumed, are of bone and ivory. A good example of the single-edged or "backed" comb is given on fig. 469; they varied much in ornamentation.

Fig. 469.

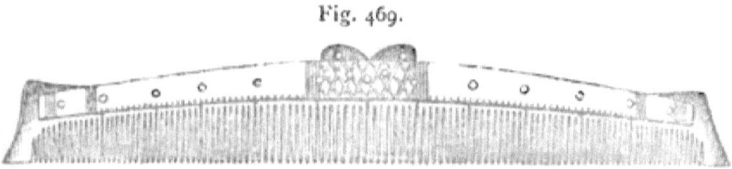

The next (fig. 470) is toothed on both its edges, and has guards or covers to fit on the teeth, in the same manner as

Fig. 470.

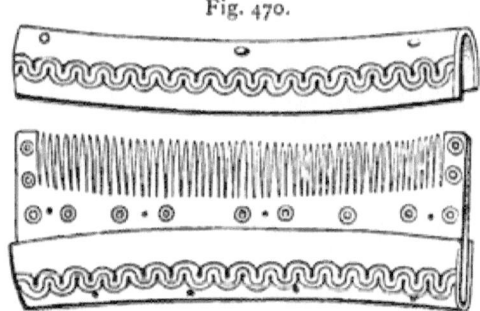

common pocket-combs of the present day. The next is a comb with a handle, which was dredged up out of the river Thames. The period is somewhat uncertain, but I give it for the purpose of comparison, as I do also the three next figures, the first of which is from the mummy graves at Arica, the second a modern wooden comb from the same district, and the third an Indian scalp-comb. Combs from Rangoon, in the Burmese empire, and from China, are also very curiously illustrative of those of early races found in our own country.

Fig. 471.

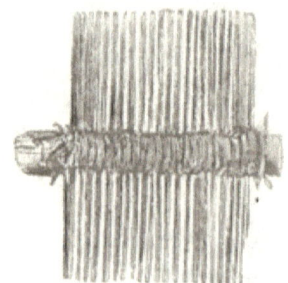

Fig. 472.

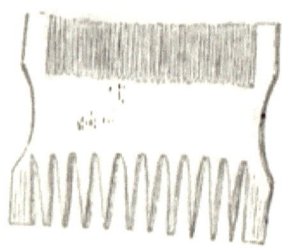

Fig. 473.

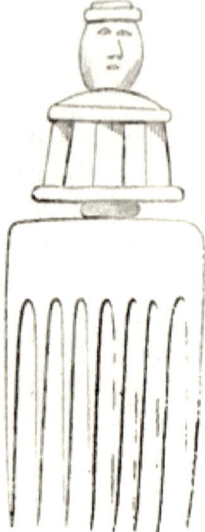

Fig. 474.

Mirrors such as are found in Roman graves are occasionally, but very rarely, met with; they were, of course, articles for the toilet. Shears or scissors of iron, some of which are of precisely the same form as our modern sheep-shears, and others of the shape of scissors of the present day, are of not unfrequent occurrence. Tweezers, too, are occasionally met with. The usual form is shown on fig. 475.

Fig. 475.

They are of bronze, and were, it is said, used for pulling out superfluous hairs from the body. They with the scissors were frequently worn attached to the girdle, along with other instruments, of which I shall now say a few words.

Châtelaines, or girdle-hangers, are among the most interesting of discoveries in the graves of Saxon females. They consist of a bunch of small implements of various kinds—keys, tweezers, scissors, tooth-picks, ear-picks, nail-cleaners, etc., and ornaments of one kind or other—hung on a chain, which being attached to the girdle hung down by the side to the thigh, or, in some instances, evidently as low as the knee. The various instruments are of silver, bronze, or iron, and are generally, the iron especially, corroded into an almost shapeless mass. The silver and bronze being more endurable, the instruments of these metals are better preserved. The example here given (fig. 476) is from one of the Kentish graves. Of some of the articles found the use is unknown, but most can be easily identified. A bunch of what is supposed to be three latch-keys is given on fig. 477, and on the next figure, 478, two curious objects, the use of which has probably been to hang small instruments on, to attach them to the girdle. For the same use, probably, are the curious and somewhat puzzling objects which are occasionally met with, and are here shown on fig. 479. They are found in pairs, attached at the top, and vary much in the pattern of the lower extremities. Probably the girdle passed through the upper part, and

Fig. 476.

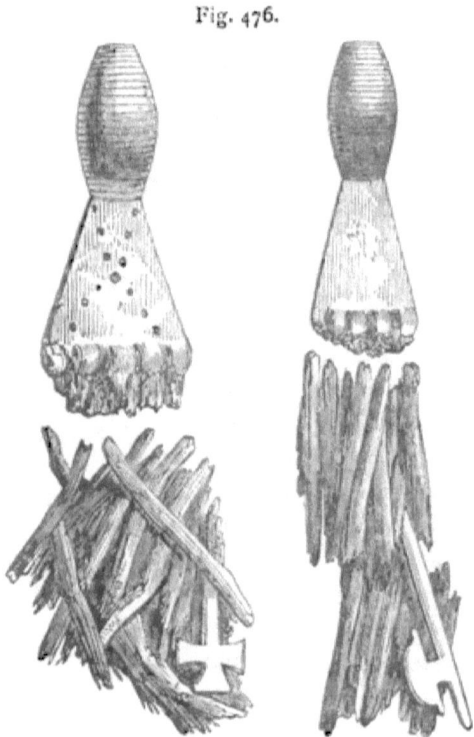

keys and other objects would be hung on the lower ends. Each side of the one here engraved is six and a half inches in length. A large variety of girdle ornaments have been found in different districts.

Hair-pins are of various forms and lengths. They are generally of bronze, but sometimes of bone. They are

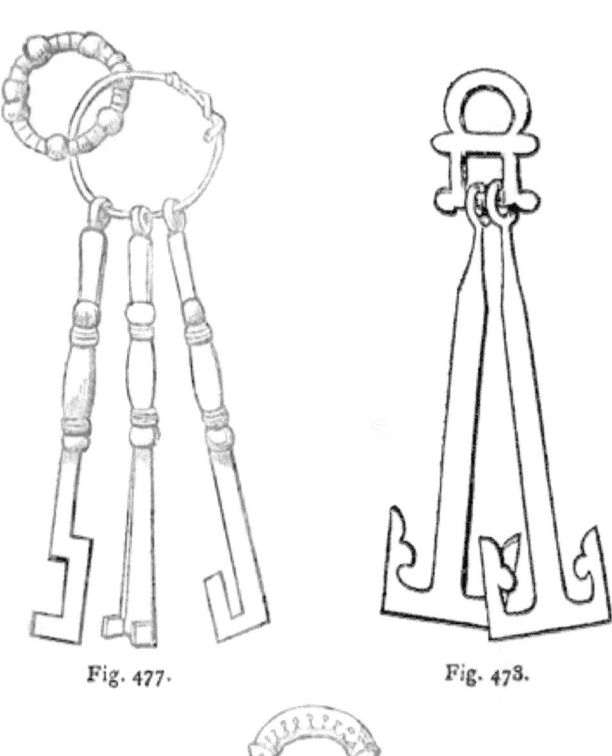

Fig. 477.   Fig. 478.

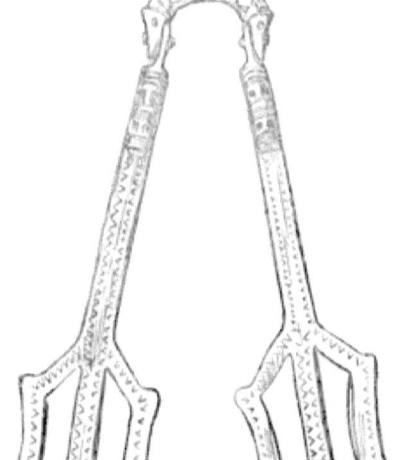

Fig. 479.

sometimes plain, but at others highly ornamented, occasionally being richly enamelled. Fig. 480 is of unique form, and has three flat pendants of bronze attached to its head by a ring. Besides hair-pins, numbers of metal pins for domestic purposes are met with.

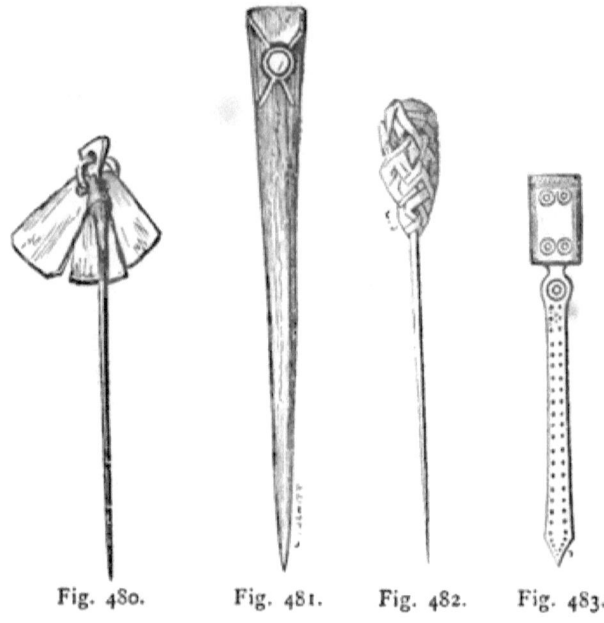

Fig. 480.   Fig. 481.   Fig. 482.   Fig. 483.

Of locks and keys, scales and weights, and many other articles, it will not be necessary to speak at further length than simply to note that they are sometimes found in Saxon graves. Bells—small hand-bells—too, are found in the graves of women. They are of bronze or iron, and of the rectangular form so characteristic of Saxon bells of larger size.

One of the most curious set of objects which the Saxon graves of Derbyshire have produced is a set of twenty-eight bone counters, or draughtmen, some of which are shown on the following engraving (fig. 484) where they are represented of their full size. They were found by Mr. Bateman in a barrow near Cold Eaton, along with an interment of burnt bones, some fragments of iron, and portions of two bone combs. The draughtmen, as they are supposed to be, and the combs, had been burnt with the body. The following is Mr. Bateman's account of this curious discovery:—

"The barrow was about twenty yards across, with a central elevation of eighteen inches, and was entirely composed of earth. The original deposit was placed in a circular hole, eighteen inches in diameter, sunk about six inches in the stony surface of the land on which the barrow was raised, so that the entire depth from the top of the latter was two feet. The interment consisted of a quantity of calcined human bones, which lay upon a thin layer of earth at the bottom of the hole, as compactly as if they had at first been deposited within a shallow basket or similar perishable vessel. Upon them lay some fragments of iron, part of two bone combs, and twenty-eight convex objects of bone, like button-moulds.

"The pieces of iron have been attached to some article of perishable material; the largest fragment has a good-sized loop, as if for suspension. One of the combs has been much like the small-tooth comb used in our nurseries, and is ornamented by small annulets cut in the bone; the other is of more elaborate make, having teeth on each side as the former, but being strengthened by a rib up the middle of both sides, covered with a finely cut herring-bone pattern, and attached by iron rivets.

"The twenty-eight bone objects (of which nine are engraved on fig. 484) consist of flattened hemispherical pieces, mostly with dots on the convex side; in some, dots within annulets. They vary from half an inch to an inch in diameter, and have generally eight, nine, or ten dots each;

but these are disposed so irregularly that it would be difficult to count them off-hand, which leads to the conclusion that these counters would not be employed for playing any game dependent upon numbers, like dominoes or dice, but that they were more probably used for a game analogous to draughts. This is most likely to be the fact, as draughtmen have occasionally been found in Scandinavian grave-

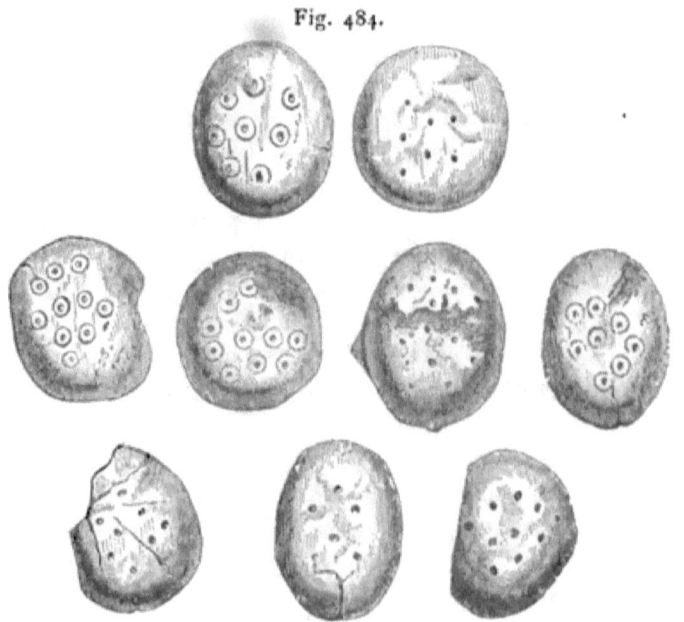

Fig. 484.

mounds; and we must assign this interment to the Saxons, whose customs were in many respects identical. All the articles found in this barrow have undergone the process of combustion, along with the human remains."

In Yorkshire, some years ago, a stone, marked in small squares like a draught-board, was found at Scambridge.*

* "Ten Years' Diggings," p. 231.

In a grave at Gilton, in Kent, two small dice, here engraved of their full size (fig. 485), were found. They were formed of ivory or bone.

Fig. 485.

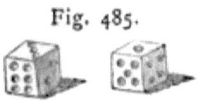

Querns, or hand-mills, for grinding corn, have on many occasions been found in or about Anglo-Saxon interments. The one engraved on the next figure (fig. 486) was found

Fig. 486.

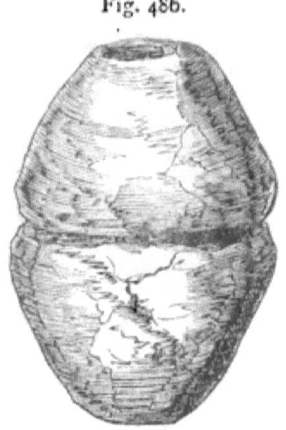

in a Saxon grave in the grounds of Miss Worsley, at Winster, along with many other interesting relics. One half of the quern had been burnt along with the body, as had also many of the stones which formed the mound.

The next (fig. 487) is from Kings Newton, the same locality referred to under the head of Anglo-Saxon pottery. Portions of stones which have evidently formed triturating stones, or grinders, are occasionally found in the grave-mounds of different periods. These have doubtless been of the same general character with the two here engraved for

Fig. 487.

comparison (figs. 488 and 489). Similar stones are found in Ireland.

Fig. 488.

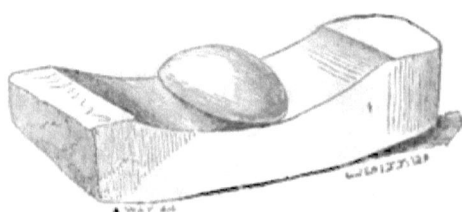

Fig. 489.

Besides the objects here spoken of, a large variety of interesting remains of a miscellaneous character are found in the Saxon graves, but which, however interesting they may be, do not require in my present work to be specially noted.

I HAVE endeavoured in the foregoing pages to give, in as brief a form as was consistent with a clear description of the objects, a faithful picture of the endless stores of treasures which the grave-mounds of our earliest forefathers open out to us, and to point out, with the aid of illustrations, the characteristics of each of the three great divisions, so as to enable my readers correctly to appropriate any remains which may come under their notice. I have purposely, and studiously, avoided theory and conjecture as far as was at all possible; contenting myself rather with bringing forward facts, which observations, personal or otherwise, into the grave-mounds and their contents have established, than speculating upon matters which can have no real bearing upon the subject.

It is said that "there is nothing new under the sun." The researches which have been made into the grave-mounds of the three great periods—the Celtic, the Romano-British, and the Anglo-Saxon—tend immeasurably to show the approximate truth of this adage, and my readers, from the foregoing pages, will be able to judge pretty correctly how many of our so-called *modern* inventions and appliances were common to, and in use by, our predecessors of "centuries and tens of centuries" of years gone by.

INDEX.

Abney Moor, 75
Abury, 71
Adzes, 109, et seq.
Aldborough, 145
Ale Glasses, 229, et seq.
Allemanic Pottery, 221
Amber Beads, 134
Anglo-Saxon Armour, 252, et seq.
,,   Arms, 236 to 264
,,   Banquet, 231
,,   Buckets, 280 to 282
,,   Cellarer, 230
,,   Coins, 235
,,   Cup-bearer, 230
,,   Fibulæ, 267 to 279
,,   Glass, 228 to 235
,,   Horse-shoes, 264
,,   Interments, 202 to 213
,,   Interments in Celtic Barrows, 13
,,   MSS. 230, 239, 240, 282
,,   Period, 202 to 298
,,   Personal Ornaments, 233 to 235
,,   Poem of Beowulf, 206, et seq., 241, 255, 256
,,   Population, 202, et seq.
,,   Pottery, 214 to 227
Animal Bones, 23, 39
Arbor-Low, 3, 71, 82, 4, 50, 71, 82, 117.
Arica, 287.
Armlets, 196, 274
Armour, 248, et seq., 253, 254
Arrow-heads, Bronze, 193

Arrow-heads, Flint, 115, et seq.
Artis, 152, et seq.
Ash, 241
Ashborne, 250
Ashbury, 67
Avisford, 147
Axe-heads, 109, et seq.

Balidon Moor, 87
Ballynageerah, 62
Banquet, 230
Barlaston, 258 to 263
Barrows, distribution of, 1
,,   Elliptical, 6
,,   Long, 5
,,   meaning of, 4
,,   (see Grave-mounds)
,,   Twin, 5
Bartlow, 142, 147, 185
Basin, Stone, 60
Baslow, 4, 33
Bateman, T., 12, 21, 115, 123, 209, 233, 250, 256, 263, 279, 293
Bath, 149
Battley, 162
Beads, Amber, 234
,,   Clay, 233
,,   Glass, 186, 187
,,   Jet, 123, et seq., 233 to 235
Bells, 292
Benty Grange, 211, 250, et seq., 257, 261
Beowulf, 206, et seq., 230, 240, 241, 255, 256
Berkshire, 67
Berriew, 80
Bishopstoke, 144
Blake-Low, 4

Blind-Low, 4
Boar-Low, 4
Boar, Sacred, 253 to 257
Bone Implements, 42, 48, 124 to 128
Bonsall, 275
Borther-Low, 4
Bottles-Low, 4
Boulders, 33
Bourne Park, 280
Bowls, 283, 284
Boxes, Bronze, 257, 284, 285, 286
Boyne, 57
Brassington Moor, 74
Brennanstown, 63
Brier Low, 3
Briggs, J. J., 225
Bronze Bows, 283, 284
" Box, 257, 284, 285, 286
" Celts, 25, 128 to 131
" Daggers, 48, 130, 131, 132
" Pins, 35
Brooch (see Fibulæ)
Broseley, 165, 168
Broughton, 35
Brown-Low, 4
Buckets, 280, 281, 282
Buckles, 218, 249, 250
Burnt Bones, 31 to 43
Buxton, 3, 122, 123, 250

Caerleon, 149
Calais Wold, 116, 120, 124
Caldon-Low, 4
Cal-Low, 4
Calver-Low, 4
Cambridgeshire, 35, 289
Carvoran, 148
Casking-Low, 4
Castleford, 196
Castor 164
" Potter's Kiln, 152
" Pottery, 152 to 162
Cellarer, 230
Celtic or Ancient British Period, 6 to 133
" Bone Articles, 123 to 126
" Bronze Celts, 128 to 132
" " Daggers, 132, 133
" Chambered Tumuli, 50 to 71
" Coins, 132, 133

Celtic Cromlechs, 27, 50 to 71
" Flint Implements, 114 to 121
" Gold Articles, 132
" Interments, 6 to 49
" Jet Articles, 122 to 125
" Pottery, 83 to 107
" Stone Circles, 10, 71 to 82
" Stone Implements, 108 to 114
Celts, Bronze, 128 to 131
" Flint, 122
" Stone, 109, 110, 111
Cemeteries, Roman, 134, et seq.
" Kingston, 212
" Kings Newton, 212, 222, et seq.
" Saxon, 212, et seq.
Chain-work, 254
Chambers, Sepulchral, 146
Chambered Tumuli, 55 to 71
Chambers of Stone, 27, 50, 55 to 71, 146, et seq.
Channel Islands, 63
Châtelaines, 289
Chatham, 160, 162
Chelmorton-Low, 4
Chester, 147
Chesters, 149
Chestersovers, 219
Chesterton, 262
Chest, Stone, 143, et seq.
Chisels, 109
Chun Cromlech, 53
Church Sterndale, 284
Cinerary Urns, Anglo-Saxon, 214, et seq.
" " Celtic, 31, 34, 84 to 95
" " Frankish, 221
" " Romano-British, 161, et seq.
Circles of Stone, 10, 71 to 82
" " (see Stone Circles)
Cist, Stone, 11, et seq., 36
Clay Coffins, 145
Cloth, Burial in, 35
" Interment in, 35, 45
" Woollen Garment, 45, 46
Cochét, 221
Cock-Low, 4
Coffins, Clay, 145

Coffins, Lead, 144
  ,,   Stone, 143
  ,,   Tile, 147
  ,,   Wood, 143
Coins, Ancient British, 133
  ,,   Roman, 32, 55, 136, 141, 187, 188
  ,,   Saxon, 235
Colchester, 143, 144, 146, 147, 157, 159, 185, 201
  ,,   Vase, 159
Cold Eaton, 293
Combs, 201, 286, 287, 288, 293
Contracted Positions, 11, et seq.
Cop-Low, 4
Cornwall, 2, 51, 75
Counters, 292 to 295
Cow Dale, 123
Cow-Low, 4, 228
Craike Hill, 43
Cremation, Interments by, 11, 31, 134, et seq., 202, et seq.
Cromlech, Ballynageerah, 62
  ,,   Brennanstown, 63
  ,,   Chun, 53, 54
  ,,   De Tus, 27
  ,,   Drumloghan, 61
  ,,   Gaulstown, 62
  ,,   Gib Hill, 43
  ,,   Glencullen, 63
  ,,   Howth, 63
  ,,   Kells, 61
  ,,   Kilternan, 63
  ,,   Kits Coty House, 53
  ,,   Knockeen, 61
  ,,   Knock Mary, 63
  ,,   L'Ancresse, 63
  ,,   Lanyon, 51, 52
  ,,   Minning-Low, 54, 55
  ,,   Molfra, 54
  ,,   Monasterboise, 61
  ,,   Mount Brown, 63
  ,,   ,, Venus, 63
  ,,   Plas Newydd, 54, 55
  ,,   Rathkenny, 63
  ,,   Shandanagh, 63
  ,,   Zennor, 54
Cronkstone-Low, 4
Cross, 253, 269
Cup-bearer, 239

Daggers, Bronze, 130, 131, 132

Daggers, Flint, 117, et seq.
  ,,   Iron, 242, 243
Danish interments, 44 to 50
Darley Dale, 92, 94
Dars-Low, 4
Dartmoor, 75
Darwen, 90
Davis, Dr., 16, 22
Derbyshire Barrows, 2, 3, 4, 16, et seq.,
Devonshire, 75
Dewlish, 7
Dice, 294, 295
Discs, enamelled, etc., 260 to 264
Dominoes, 294
Dorsetshire Barrows, 2, 3, 7, 47, 91
Double interments, 25, 29, 30
Dove Dale, 128
Dow-Low, 4
Dowth, 59, 61, 66
Drake-Low, 4
Draughtboard, 294
Draughtmen, 292, 293, 294
Draughts, Game, 292, 293, 294
Drinking Cup, 43, 44, 100 to 104, 251, 282, 283
Druidical Circles, 10, 71 to 82
Durobrivian Pottery, 152 to 162

Earl Stemdale, 3
Ear-picks, 289
East-Low Hill, 146
East-Moor, 75
Elk-Low, 4, 72
Ely, 107
Enamelled Discs, etc., 260 to 264
Enamels, 251, 266, 267, et seq.
  ,,   Chinese, 260
  ,,   Roman, 196
  ,,   Saxon, 260 to 264
End-Low, 4
Envermeu, 282
Extended positions, 11, et seq.

Fairford, 280
Fairholt, F.W., 278
Farlow, 4
Faussett Collection, 217
Fibulæ, Anglo-Saxon, 266 to 279
  ,,   Roman, 193 to 196
Fimber, 43, 44, 97, 124
Flax Dale, 33, 71

INDEX.

Fleming, G., 264
"Flint-Jack," 115
Flint Acutely Angled, 119
   „ Barbed Arrow-heads, 115, 116
   „ Celts, 122, 123
   „ Dagger-blades, 117, 118
   „ Flakes, 121,
   „ Implements, 115 to 123
   „ Leaf-shaped, 119
   „ Notched, 118, 120
   „ Thumb, 122
   „ Various, 121, 122
Food Vessels, 44, 95 to 100
Foo Low, 4
Fowse-Low, 4
Fox-Low, 4
Frankish Pottery, 221
Froggatt Edge, 75

Galley-Low, 4
Garment, Woollen, 45
Gaulstown, 62
Germany, 160
Gib-Low, 4
Gilton, 295
Girdle-hangers, 289, 290, 291
   „ Ornaments, 290, 291
Glass, Ale, 229, 230, 232
   „ Beads, 185, 231 to 235
   „ Bowls, 186, 228, 229
   „ Decanters (?) 231
   „ Lachrymatories, 186
   „ Roman, 145, 185 to 188
   „ Saxon, 225 to 235
   „ Sepulchral Vessels, 185
   „ Tumblers, 229
Glencullen, 63
Gloucester, 201
Gloucestershire, 70
Gold Articles, 132, 133, 266 to 279
   „ Drops, 279
   „ Torques, 132, 196 to 199
Gospel Hillock, 104, 121, 123, 124
Grave-mounds, Anglo-Saxon, 202 to 298
   „   „ Celtic, 6 to 132
   „   „ Construction of, 6, et seq., 33, 38, 134 to 143, 202 to 213
   „   „ Danish (?) 44 to 50
   „   „ Distribution of, 2

Grave-mounds, Romano-British, 134 to 201
Great-Low, 4
Green-Low, 4, 114, 115
Grimthorpe, 238, 245, 246, 263
Grinders (see Querns)
Grind-Low, 4, 100
Gris-Low, 4
Gristhorpe, 44
Grub-Low, 4
Gruter, 135
Guernsey, 27
Gunthorpe, 116, 120

Haddon, 141
Hair-pins, 290, 292
Hammer-head, 42, 109, et seq.
Hampshire, 143, 149
Hand-mills, 295, 296
Hard-Low, 4
Har-Low, 4
Hartington, 3
Hartle Moor, 74
Hatchet, 109, 113
Hathersage Moor, 75
Hawks-Low, 4
Hav Top, 100
Helmets, 248, et seq.
Heins-Low, 4
High-Low, 4
   „ Needham, 3
Hitter Hill, 6, et seq., 16, 98
Hob Hurst's House, 33
Hog's Bones, 23
Horning-Low, 4
Horse-shoes, 201, 264, 265
Horsley, 114
Houe, meaning of, 4
Howth, 63
Huck-Low, 4

Immolation of Infants, 106
   „   „ Slaves, 106
   „   „ Wives, 91, 106
Incense Cups, 84, 104 to 107
Inscriptions, Sepulchral, 135, 148, 149, 150
Interment by Cremation, 11, 31, 134, et seq., 202, et seq.
   „   „ Inhumation, 11 to 49, 134, et seq.

Interment in Cloth, 35, 45, 46
    „   „   Skin, 35
    „   „   Tree-Coffins, 44 to 50
    „   „   Pit, 43
Inverted Urns, 33, 34
Ireland, 28, 63, 113

Javelins, 243, 244
Jet, 25, 44, 123 to 126
 „ Necklaces, 44, 123, 124, 125
 „ Pendants, 124, 126
 „ Ring, 126
 „ Studs, 123, 124, 126
Jutland, 46

Kells, 28
Kens-Low, 4
Kent, 53
Keys, 201, 289, 292
Kilkenny, 63
Kilternan, 63
Kingsholme, 144
Kingston, 212, 215, 266, 267
Kings Newton, 212, 214 to 227, 295
Kirk Michael, 274
Kit's Coty House, 53
Kneeling position, 11, et seq.
Knives, 193, 242, 243
Kneck-Low, 4
Knok Mary, 63
Knot-Low, 4

Lady-Low, 4
Laidman's-Low, 4
Lake Dwellings, 45
Lamp, 201
Lancashire, 90
Lapwing Hill, 209
Lark's-Low, 4
Lead Coffins, 144, 145
 „ Ore, 31
 „ Pigs of, 32
 „ Smelting, 32
Lean-Low, 4
Leckhampton, 258
Lewes, 257
Liffs-Low, 4, 42
Lillebonne, 177
Lincoln, 257
Lincolnshire, 35

Lindenschmidt, 219, 265
Little Chester, 142, 168, 169, 190
Locks, 201, 292
Lollius, 135
Lomber-Low, 4
Londinières, 221
London, 135, 142, 143, 144, 148, 171, 175, et seq.
Long Low, 36
Lord's Down, 7
Low, meaning of, 4
 „ (see Grave-mounds)
Lowsey-Low, 4
Lukis, Capt. 123
 „   F. C., 27

Mail, Coat of, 255, 256
Mauls, 109, et seq.
Mayence, 219
May-Low, 4
Medway, 160
Mick-Low, 4
Mickleover, 114
Middleton, 3, 33, 41, 123, 261
Minning-Low, 54, 141
Mirrors, 199, 290
Modelling Tools, 124
Money-Low, 4
Monsal Dale, 28, 86, 98
Mortimer, 43, 44, 97, 124
Moot-Low, 4, 127, 128
Mount Brown, 63
 „ Venus, 63
Musden-Low, 4
Mutti-Low Hill, 35

Nail-cleaners, 289
Necklace, Glass, 187, 232, et seq.
 „   Jet, 44, 123 to 126
 „   „ and Bone, 124
Needham-Low, 4
Needwood, 198
Nen, 152
Nether-Low, 4
New Forest, 149, 165
 „ Grange, 61, 66
Newhaven, 3, 256
Normandy, 174
North Elmham, 217
Northumberland, 46, 148
Nowth, 59

Ochre, 43
Off-Low, 4
Oghams, 61, 277
Otterham Creek, 162
Over Haddon, 284
Oxfordshire, 164
Ox-Low, 4
Ozengall, 144, 211

Painstor-Low, 4
Palstaves, 128
Paradise, 27
Parcelly Hay, 3, 25, 26
Pars-Low, 4
Parwich, 141
Peg-Low, 4
Pendants, Bone, 125, 126
,, Enamelled, etc., 260 to 264
,, Gold, 279
,, Jet, 124 to 126
Penannular Brooch (see Fibulæ)
Phœnix Park, 63
Pigtor-Low, 4
Pike-Low, 4
Pinch-Low, 4
Pins, Hair, 290, 292
Pit Interments, 43, 44
Plymouth, 192, 193, 199
Pottery, Amphoræ, 171, 172
,, Anglo-Saxon, 214 to 227
,, Celtic, 83 to 108
,, Domestic Vessels, etc., 170 to 174
,, Drinking Cups, Celtic, 100 to 104
,, Durobrivian or Castor, 151, 152 to 162
,, Food Vessels, Celtic, 95 to 100
,, Frankish, 214 to 227
,, Hampshire, 151, 165, 166
,, Handled Cups, Celtic, 107
,, "Incense Cups," Celtic, 104 to 107
Potters' Kilns, 152, 154, 183
,, Marks, 176, 177, 178
,, Mortaria, 172, 173
,, Punches, 227
,, Sepulchral Urns, Celtic, 31, 34, 84 to 95

Potters' Sepulchral Urns, Roman, 156 et seq.
,, Sepulchral Urns, Saxon, 215 to 227
,, Stamps, 177, 227
,, Unguentaria, 171 to 174
Potters, Manufacture of, 84, 152 to 184, 227
,, Romano-British, 151 to 184
,, Salopian, 151, 164, 165
,, Samian, 151, 157 to 184
,, Upchurch, 151, 162, 163, 164
,, Yorkshire, 151, 166

Queen-Low, 4
Querns, 295, 296 (see also Grinders and Triturating Stones)

Rains-Low, 4
Rangoon, 287
Rats' Bones, 16, 87, 90
Ravens-Low, 4
Red Ochre, 43
Repton, 213
Ribden-Low, 4
Rick-Low, 4
Rigollot, 220
Ringham-Low, 4, 116, 119, 120
Rings, 235
,, Jet, 124, 126
Rochester, 147
Rocky-Low, 4
Rollrich, 71
Rolly-Low, 4, 34
Roman Arms, etc., 190, et seq.
,, Cemeteries, 134, et seq.
,, Coins, 55, 141, 187, 188
,, ,, as payment for passage over Styx, 136, 141
,, Glass 184 to 188
,, Personal Ornaments, 193, et seq.
,, Population, 134, et seq.
,, Pottery, 151 to 184
Romano-British Period, 134 to 201
Rouge, 43
Round-Low, 4, 32
Roundway Hill, 16, 100
Rusden-Low, 4
Runes, 241
Sacrifice of Infants, 106

INDEX.

Sacrifice of Slaves, 106
,, Wives, 91, 106
Saint-Low, 4
Salona, 147
Salopian Pottery, 164, 165
Samian Ware, 175 to 184
Sancreed, 76
Sarcophagus, 143, et seq.
Scales and Weights, 292
Scambridge, 294, 295
Scarborough, 47
Scissors, 289
Scrapers of Flint, 121
Seax, 240, et seq.
Selzen, 219
,, 265
Sepulchral remains, Anglo-Saxon, 202 to 298
,, ,, Celtic, 1 to 133
,, ,, Frankish, 221
,, ,, Danish, 44 to 50
,, ,, Romano-British, 134 to 201
,, Chambers, 146
,, Glass, 185
,, Inscriptions, 135, 148, et seq., 217
,, Urns (see Cinerary Urns)
Shandanagh, 63
Shears, 289
Shields, 243 to 248
,, Umbones of, 246, 247, 261
,, from MSS., 248
Shuttlestone-Low, 24, 130
Sibertswold, 247, 282
Sitting-Low, 4
Sitting position, 11, et seq.
Skeleton, positions of, 11, et seq.
Skins, interment in, 24, 35
Skull, Hitter Hill, 21
,, distributions of, 22
,, Long-Low, 39
,, Gristhorpe, 47
Sliper-Low, 5
Smerrill Moor, 12
Smith, C. R., 160, 164, 204, 216, 255
Southfleet, 144
Spear-heads, 190, 192, 243, 244
Spindle-whorls, 114
Staden-Low, 4
Staffordshire Barrows, 4, 86, 89, 92, 96

Stan-Low, 4
Stanshope, 132
Stanton Moor, 73
Sterndale, 33, 284
Stone Chambers, 27, 50, 55 to 71, 146, et seq.
Stone Circles, 10, 27, 34. 71 to 82
,, ,, Abney Moor, 75
,, ,, Abury, 71
,, ,, Arbor-Low, 3, 71, 82
,, ,, Berriew, 80
,, ,, Boscawen-Un, 80
,, ,, Brassington Moor, 74
,, ,, Channel Islands, 78
,, ,, Cornish, 75
,, ,, Dartmoor, 75
,, ,, East Moor, 75
,, ,, Elk-Low, 72
,, ,, Eyam Moor, 74
,, ,, Flax Dale, 71
,, ,, formation of, 71
,, ,, Froggatt Edge, 75
,, ,, Hartle Moor, 74
,, ,, Hathersage Moor, 75
,, ,, Isle of Man, 76, 78
,, ,, Mule Hill, 78
,, ,, "Nine Ladies," 73, 74
,, ,, Penmeanmaur, 80, 81
,, ,, Rollrich, 71
,, ,, Sancreed, 76
,, ,, Stanton Moor, 73, 74
,, ,, Stonehenge, 71
,, ,, Trewavas, Head, 76
Stone Cists, 11, 17, et seq., 33, 36, et seq., 143, et seq.
,, Coffins, 143, 144, et seq.
,, Implements of, 109, et seq.
Stone, 92
Stoney Littleton, 67
Stonehenge, 371
Stowborough, 47
Strigils, 201
Studs, Bone, 122, 126
,, Jet, 124, 126
Sussex, 146
Suttee, 91
Sutton Brow, 92

INDEX

Swinscoe, 22
Swiss Lake Villages, 45
Swords, Roman, 190, 191
　,,　　Saxon, 236 to 242
　,,　　from MSS., 239, 240
Swordsman, 240

Taddington, 67, 69
Tara Brooch, 278
Thirkel-Low, 4
Thirsk, 92
Thoo-Low, 4, 5
Three-Lows, 5
Thumb Flints, 121
Tile Tombs, 147, 148
Tissington, 13, 211, 236, 247
Toothpicks, 289
Torques, 133, 196 to 199
Totmans-Low, 4
Tree-Coffins, 44, 45, 50
Trentham, 89, 96
Triturating Stones, 114, 295, 296
　(see also "Querns")
Tump, meaning of, 4
Tumuli, Chambered, 55 to 71
　(see Grave-mounds)
Tumulus, Etruscan, 55
Twin-Barrows, 37, 78, 79
Tweezers, 201, 289

Uley, 70
Umbones of Shields, 246, 247, 261
Upchurch, 162, et seq.
　,,　　Pottery, 162 to 164
Upright position, 11, et seq.
Uriconium, 137, (see also Wroxeter)

Vale, 27
Vole, Water, 16, 89, 90

Ward-Low, 5, 34
Warry-Low, 5
Water Rat, 16, 89, 90
　,,　　Vole, 16, 89, 90
Wath, 47
Wedgwood, F., 258
Weights, 292
Wellbeloved, 163
Wellow, 67
West Lodge, 157
Westwood, 253
Wetton, 193
Whetstones, 114
White-Low, 5
Willoughby, 113
Wilson, C. C., 278
Wiltshire Barrows, 2, 16, 100
Winster, 3, 111, 211, 268, 269, 295, 296
Withery-Low, 5
Woolaton, 109
Woollen Cloth, 45
Wool-Low, 5
Worsaae, 255
Worsley, Miss, 295
Wright, T., 135, 151, et seq., 176, et seq., 216
Wroxeter, 137, 141, 147, 162 to 165
Wyaston, 210, 233
Wye, 28
Wykeham, 98

Yarns-Low, 5
York, 142, 143, 144, et seq.
Yorkshire Barrows, 2, 5, 7, 25, 35, 44, 47, 97, 164
　,,　　Pottery, 151
Youlgreave, 33

Dedicated to the Right Hon. LORD LYTTON.

In One handsome Volume, Foolscap 4to., cloth gilt, price 25s.

WOMANKIND
IN WESTERN EUROPE,
From the Earliest Ages to the Seventeenth Century.

By THOMAS WRIGHT, M.A., F.S.A.

Illustrated with numerous Coloured Plates and Wood Engravings.

"It is something more than a drawing-room ornament. It is an elaborate and careful summary of all that one of our most learned antiquaries, after years of pleasant labour, on a very pleasant subject, has been able to learn as to the condition of women from the earliest times. It is beautifully illustrated, both in colours—mainly from ancient illuminations—and also by a profusion of woodcuts, portraying the various fashions by which successive ages of our history have been marked."—*The Times.*

"We should be at a loss to find words of excessive praise for the learning, judgment, and delicate art with which the author has gathered, arranged, and presented the multifarious materials of a fascinating narrative, that would be told effectively by the embellishments of the book, even if the illustrations were not accompanied with words of explanatory text."—*Athenæum.*

"This is much more than a pretty illustrated book. It is a repertory of antiquarian literature on the costume, social habits, domestic pursuits, and position of the sex, and the illustrations are from all sorts of recondite sources—MS. illuminations of the Romances, Psalters, and Chronicles. It reflects great credit on the writer, whose vast stores of information and research have been, in this instance, well employed. The volume is quite an encyclopædia on a special subject."—*Saturday Review.*

"As a work of art, no less than of literary elucidation, this book is perfect in all its parts, and most honourable to its publishers. . . . The letterpress enhances the value of the work itself a hundredfold, as might have been expected from so well known and learned an antiquarian as Mr. Wright, whose participation in so choice a work makes it in every respect worthy of a place in every public and well-selected library, where art and literature are alike patronized and admired."—*Bell's Weekly Messenger.*

"We cannot justly class Mr. Wright's 'Womankind' amongst the ephemeral books of the season; yet it is admirably suited to answer the purpose of a gift-book—and much more; and it would be unfair to leave it until its less solid neighbours had been cleared out of hand.

The high antiquarian renown of the author would alone guarantee that we should have no frivolous, superficial dissertation on the mere outward phenomena of 'feminity' in past times—no mere sentimental declamation in favour of woman's advancement to a social place which she never before claimed. On the contrary, we have a faithful, unshrinking, photographically minute account of the relations between women and men, and of female manners, dress, social duties, and positive literary achievements, and participation in public life, from the date at which authentic history takes cognizance of the condition of the European nations. . . . Mr. Wright's 'Womankind'—like the ideal of the gentle sex—is fitted, not for the festive season alone, but for every time.—*Daily Telegraph*.

"The author's name, on whatever subject he writes, is a guarantee for thorough scholarship, solid information, lucid exposition, and careful delineation; and in this work all these qualities are conspicuous. Mr. Wright believes, and with good reason, 'that a history of the female sex, in that particular division of mankind to which we ourselves belong, would not be unacceptable to the general reader.' Such a history he has here produced, and in doing so, has left nothing to be desired. . . . In every sense this is a splendid book, for which we heartily thank Mr. Wright."—*Illustrated Times*.

"Never has history been made more charming than in this excellent volume. Whatever page is opened, some pleasant little narrative, historic or romantic: some sketch of the womankind of Chaucer's days, or of the heroines of the Romaunt of the Rose; some striking pictures of Anglo-Saxon life, or some quaint costumes, or ever-changing fashions, constantly attract, and interest, and inform."—*Birmingham Daily Post*.

"To the general public, the appearance of such a work is a surprise, the more agreeable because, while it is the work of an accomplished scholar, who has nowhere deviated from the scholar's path to win ephemeral applause, it nevertheless appeals to universal sympathies, and so abounds in attractions as to demand to be regarded as emphatically *the* book of the season."—*Gardeners' Magazine*.

"Externally and internally it is absolutely splendid, the binding and illustrations being a perfect marvel of beauty and richness. But in the interest of its subject, as well as in its mode of treatment, Mr. Wright's present work will command the respect and praise of the man of letters and the philosopher, quite as much as it is sure to enlist the sympathies and extort the admiration of a less exacting class of readers. The book is beautifully written, the style being at once chaste and ornate."—*Eddowes's Shrewsbury Journal*.

"It is one of the most interesting, instructive, and valuable books of the nineteenth century. At this particular period of the agitation of woman's rights, we may say in truth that this book is a treasury of knowledge to the historian, the politician, the moral philosopher, and the reformer; while, at the same time, in its romantic incidents illustrative of social life in different ages of Western Europe, it surpasses in interest the most skilful and attractive fictions of the day."—*New York Morning Herald*.

GROOMBRIDGE & SONS, 5, PATERNOSTER ROW, LONDON.

www.ingramcontent.com/pod-product-compliance
Lightning Source LLC
Chambersburg PA
CBHW021155230426
43667CB00006B/415